VOLUME ONE HUNDRED AND THIRTY NINE

ADVANCES IN CANCER RESEARCH

Cancer Nanotechnology

VOLUME ONE HUNDRED AND THIRTY NINE

Advances in CANCER RESEARCH

Cancer Nanotechnology

Edited by

ANN-MARIE BROOME

Department of Cell and Molecular Pharmacology & Experimental Therapeutics; Hollings Cancer Center; Department of Neurosciences, Medical University of South Carolina, Charleston, SC, United States

ELSEVIER

ACADEMIC PRESS

An imprint of Elsevier

Academic Press is an imprint of Elsevier
50 Hampshire Street, 5th Floor, Cambridge, MA 02139, United States
525 B Street, Suite 1650, San Diego, CA 92101, United States
The Boulevard, Langford Lane, Kidlington, Oxford OX5 1GB, United Kingdom
125 London Wall, London, EC2Y 5AS, United Kingdom

First edition 2018

ISBN: 978-0-12-814169-4
ISSN: 0065-230X

For information on all Academic Press publications
visit our website at https://www.elsevier.com/books-and-journals

Publisher: Zoe Kruze
Acquisition Editor: Zoe Kruze
Editorial Project Manager: Shellie Bryant
Production Project Manager: James Selvam
Cover Designer: Greg Harris

Typeset by SPi Global, India

CONTENTS

CONTRIBUTORS

Rebaz A. Ahmed
Department of Pathology; Stephenson Cancer Center; Graduate Program in Biomedical Sciences, University of Oklahoma Health Sciences Center, Oklahoma City, OK, United States

Narsireddy Amreddy
Department of Pathology; Stephenson Cancer Center, University of Oklahoma Health Sciences Center, Oklahoma City, OK, United States

Ann-Marie Broome
Department of Cell and Molecular Pharmacology & Experimental Therapeutics; Hollings Cancer Center; Department of Neurosciences, Medical University of South Carolina, Charleston, SC, United States

Dean M. Connor
Department of Cell and Molecular Pharmacology & Experimental Therapeutics; Hollings Cancer Center, Medical University of South Carolina, Charleston, SC, United States

Michaela Cooley
Department of Radiology, Case Western Reserve University, Cleveland, OH, United States

Marie-Christine Daniel
Department of Chemistry and Biochemistry, University of Maryland Baltimore County, Baltimore, MD, United States

Al de Leon
Department of Radiology, Case Western Reserve University, Cleveland, OH, United States

Lance Dockery
Department of Chemistry and Biochemistry, University of Maryland Baltimore County, Baltimore, MD, United States

Agata A. Exner
Department of Radiology; Department of Biomedical Engineering, Case Western Reserve University, Cleveland, OH, United States

Dieter Haemmerich
Department of Pediatrics, Medical University of South Carolina, Charleston, SC, United States

C. Tilden Hagan IV
Laboratory of Nano- and Translational Medicine, Lineberger Comprehensive Cancer Center, Carolina Center for Cancer Nanotechnology Excellence, Carolina Institute of Nanomedicine; Department of Radiation Oncology, Lineberger Comprehensive Cancer Center; UNC/NCSU Joint Department of Biomedical Engineering, University of North Carolina at Chapel Hill, Chapel Hill, NC, United States

Olive Jung
Department of Radiology, Case Western Reserve University, Cleveland, OH, United States; Nuffield Department of Medicine, University of Oxford, Oxford, United Kingdom

Linmei Li
Dalian Institute of Chemical Physics, Chinese Academy of Sciences, Dalian, China

Bingcheng Lin
Dalian Institute of Chemical Physics, Chinese Academy of Sciences, Dalian, China

Yao Lu
Dalian Institute of Chemical Physics, Chinese Academy of Sciences, Dalian, China

Yusra B. Medik
Laboratory of Nano- and Translational Medicine, Lineberger Comprehensive Cancer Center, Carolina Center for Cancer Nanotechnology Excellence, Carolina Institute of Nanomedicine; Department of Radiation Oncology, Lineberger Comprehensive Cancer Center, University of North Carolina at Chapel Hill, Chapel Hill, NC, United States

Anjan Motamarry
Department of Pediatrics; Department of Drug Discovery and Biomedical Sciences, Medical University of South Carolina, Charleston, SC, United States

Anupama Munshi
Stephenson Cancer Center; Department of Radiation Oncology, University of Oklahoma Health Sciences Center, Oklahoma City, OK, United States

Pinunta Nittayacharn
Department of Biomedical Engineering, Case Western Reserve University, Cleveland, OH, United States

Reshani Perera
Department of Radiology, Case Western Reserve University, Cleveland, OH, United States

Rajagopal Ramesh
Department of Pathology; Stephenson Cancer Center; Graduate Program in Biomedical Sciences, University of Oklahoma Health Sciences Center, Oklahoma City, OK, United States

Mohammad Razaq
Stephenson Cancer Center; Department of Medicine and Hematology Oncology, University of Oklahoma Health Sciences Center, Oklahoma City, OK, United States

Shiqun Shao
Department of Chemistry, University of California Riverside, Riverside, CA, United States

Qihui Shi
Minhang Branch, Zhongshan Hospital and Institute of Biomedical Sciences; Institute of Fudan-Minhang Academic Health System, Minhang Hospital, Fudan University, Shanghai, China

Akhil Srivastava
Department of Pathology; Stephenson Cancer Center, University of Oklahoma Health Sciences Center, Oklahoma City, OK, United States

Rheal Towner
Department of Pathology, University of Oklahoma Health Sciences Center; Advanced Magnetic Resonance Center, Oklahoma Medical Research Foundation, Oklahoma City, OK, United States

Andrew Z. Wang
Laboratory of Nano- and Translational Medicine, Lineberger Comprehensive Cancer Center, Carolina Center for Cancer Nanotechnology Excellence, Carolina Institute of Nanomedicine; Department of Radiation Oncology, Lineberger Comprehensive Cancer Center, University of North Carolina at Chapel Hill, Chapel Hill, NC, United States

Min Xue
Department of Chemistry, University of California Riverside, Riverside, CA, United States

Shiqiang Yan
Dalian Institute of Chemical Physics, Chinese Academy of Sciences, Dalian, China

Yan Daniel Zhao
Stephenson Cancer Center; Department of Biostatistics and Epidemiology, University of Oklahoma Health Sciences Center, Oklahoma City, OK, United States

PREFACE

The Hollywood science fiction movie from the mid-1960s, *Fantastic Voyage*, presented an intriguing tale that seemed to portend the future of nanotechnology—a miniaturized submarine, the Proteus, is used to cruise human blood vessels in an attempt to rescue the protagonist from dire damage to his brain as he lies in a coma. After overcoming many harrowing obstacles, the crew, utilizing a localized laser to bust the clot, cures the scientist; the abandoned submarine is attacked by the immune system; and the crew is summarily excreted from the scientist's body just in the nick of time. But if the word "miniature submarine" is replaced by the term "nanoparticle," this impossible movie might have, in fact, been an inspiration for the development of nanomedicine and personalized healthcare.

In this way, nanotechnology is constantly evolving and covers many areas of science, research, and technology. In the broadest sense, it is the intersection of physics, chemistry, and biology. Modern nanotechnology combines science and art to manipulate and rearrange individual atoms and molecules to create useful materials, devices, and systems. At one-billionth of a meter, nanoparticles are being developed that bend the conventional laws of physics. As a result, the properties of nanoparticles are different at the nanoscale. First, nanoparticles have a relatively large surface area when compared to the same mass of material produced in a larger form with more atoms or molecules near the surface which makes them more chemically active. Second, quantum size effects begin to dominate the behavior of matter at the nanoscale and this has consequences for optical, electrical, and magnetic behaviors.

With the use of nanotechnology in medicine (nanomedicine), scientists hope to prevent illness and more quickly diagnose, control, and treat disease with fewer side effects. Current trends in nanoparticle research parallel these goals and concentrate on issues associated with drug delivery, therapeutic development, and multifunctional, controllable drug carriers. Advantages of using nanoparticles in medicine are threefold: (1) nanoparticles smaller than 100 nm can easily enter cells; (2) nanoparticles smaller than 20 nm can transit out of blood vessels; and (3) nanoparticles are capable of holding thousands of small molecules, i.e., drugs or contrast agents. As a result, most nanoparticles used in nanomedicine tend to be sized between 1 and 500 nm. Targeted strategies for nanoparticles also overcome many of the problems associated with systemic drug delivery: lower drug doses, greater control over toxicity and bioavailability of the drug, extended duration of release, controlled release directly to the target organ, tissue or cell, and avoidance of systemic drug exposure. The application of nanoparticles in medicine is playing an ever-increasing role in cancer care. The aim of this book is to introduce several different types of nanoparticles and their novel utility in several different cancer types. It will examine the increasing complexity of nanoparticles and improved outcomes for nanoparticle delivery, detection, and disease treatment.

ANN-MARIE BROOME, PhD, MBA
May, 2018

CHAPTER ONE

Exosomes as Theranostics for Lung Cancer

Akhil Srivastava[*,†], Narsireddy Amreddy[*,†], Mohammad Razaq[†,‡], Rheal Towner[*,§], Yan Daniel Zhao[†,¶], Rebaz A. Ahmed[*,†,‖], Anupama Munshi[†,#], Rajagopal Ramesh[*,†,‖,1]

[*]Department of Pathology, University of Oklahoma Health Sciences Center, Oklahoma City, OK, United States
[†]Stephenson Cancer Center, University of Oklahoma Health Sciences Center, Oklahoma City, OK, United States
[‡]Department of Medicine and Hematology Oncology, University of Oklahoma Health Sciences Center, Oklahoma City, OK, United States
[§]Advanced Magnetic Resonance Center, Oklahoma Medical Research Foundation, Oklahoma City, OK, United States
[¶]Department of Biostatistics and Epidemiology, University of Oklahoma Health Sciences Center, Oklahoma City, OK, United States
[‖]Graduate Program in Biomedical Sciences, University of Oklahoma Health Sciences Center, Oklahoma City, OK, United States
[#]Department of Radiation Oncology, University of Oklahoma Health Sciences Center, Oklahoma City, OK, United States
[1]Corresponding author: e-mail address: rajagopal-ramesh@ouhsc.edu

Contents

Abstract

Extensive research in genetics and genomics has revealed that lung cancer is a physiologically complex and genetically heterogeneous disease. Although molecular targets that can yield favorable response have been identified, those targets cannot be exploited due to the lack of suitable drug carriers. Furthermore, lung cancer often is

Advances in Cancer Research, Volume 139
ISSN 0065-230X
https://doi.org/10.1016/bs.acr.2018.04.001

diagnosed at an advanced stage when the disease has metastasized. Conventional treatments are not effective for treating metastatic lung cancer. Targeted therapeutics while beneficial has challenges that include poor tumor-targeting, off-target effects, and development of resistance to therapy. Therefore, improved drug delivery systems that can deliver drugs specifically to tumor will produce improved treatment outcomes.

Exosomes have a natural ability to carry functional biomolecules, such as small RNAs, DNAs, and proteins, in their lumen. This property makes exosomes attractive for use in drug delivery and molecular diagnosis. Moreover, exosomes can be attached to nanoparticles and used for high precision imaging. Exosomes are now considered an important component in liquid biopsy assessments, which are useful for detecting cancers, including lung cancer. Several studies are currently underway to develop methods of exploiting exosomes for use as efficient drug delivery vehicles and to develop novel diagnostic modalities. This chapter summarizes the current status of exosome studies with regard to their use as theranostics in lung cancer. Examples from other cancers have also been cited to illustrate the extensive applicability of exosomes to therapy and diagnosis.

1. INTRODUCTION

Lung cancer is the leading cause of cancer-related death worldwide (ACS Statistics, 2018). Despite the availability of newer therapeutic methods, such as targeted therapy, immunotherapy, and combined modalities, the poor survival rates of patients with lung cancer are mainly due to late-stage diagnosis and/or the use of inefficient therapeutic regimens. An understanding of the molecular mechanisms, the ability to detect cancer at early stages, and the development of safe and highly efficient targeted therapy methods are crucial for lung cancer treatment. In this context, extracellular vehicles (EVs), especially exosomes, have received recent interest for their role in facilitating early detection and diagnosis and improving treatment outcomes in cancer (Vanni, Alama, Grossi, Dal Bello, & Coco, 2017). Accumulating evidence suggests that the genetic materials, proteins, and lipids carried by exosomes originating from cancer cells may have unique signatures that allow us to identify specific biomarkers for cancer identification and to predict therapy outcome (Kadota, Yoshioka, Fujita, Kuwano, & Ochiya, 2017).

Lung cancer-derived exosomes are known to represent the cell of origin in many aspects. A recent study demonstrated that exosomes derived from lung tumor cells (H292, H1975, and H1650) in culture reflected the mutational status of specific genes (EGFR) of the parental cell lines (Thakur et al., 2014).

In a different in vitro study, microvesicles/exosomes derived from epidermoid carcinoma (A431), lung adenocarcinoma (A549), and colorectal carcinoma (DLD-1) cell lines transferred activated EGFR from cancer cells to endothelial cells and induced the mitogen-activated protein kinase (MAPK) and AKT pathways (Al-Nedawi, Meehan, Kerbel, Allison, & Rak, 2009). Further, the study demonstrated that EGFR transfer by microvesicles resulted in the activation of vascular endothelial growth factor receptor 2 (VEGFR-2) expression in the host cell. These studies show that molecular profiling of lung cancer-derived exosomes would be a feasible technique for understanding the tumor's molecular status.

Exosomes are reported to be involved in the underlying mechanism of chemotherapy resistance in lung cancer. Cisplatin resistance is common in lung cancer. Exosomes derived from a cisplatin-resistant lung cancer cell line (A549) had variable contents, suggesting a role for exosomes in mediating cisplatin resistance (Xiao et al., 2014). Exosomes are also involved in inducing treatment resistance to concurrent chemotherapy and tyrosine kinase inhibitor. Recently, Li et al. showed that when exosomes derived from gefitinib-treated lung cancer cells and cisplatin were combined, the cisplatin sensitivity of the cancer cells was significantly reduced, whereas a combination of exosomes from cisplatin-treated cells and gefitinib had no significant effect on gefitinib-induced cell apoptosis (Li et al., 2016). When they blocked the exosome production in lung cancer cells and tested the cisplatin and gefitinib combination, a synergistic effect was observed. These findings underline the importance of exosomes in mediating the antagonistic effects of cisplatin and gefitinib therapy in lung cancer.

The peculiar double-membrane structure of exosomes resembles the structure of a commercial drug delivery vehicle, the liposome. Exosomes are engineered to carry anticancer therapeutics by active (sonication, electroporation, or freeze–thaw cycles) or passive (simple incubation with exosomes) drug-loading approaches (Luan et al. 2017). Munagala, Aqil, and Gupta (2016) and Munagala, Aqil, Jeyabalan, and Gupta (2016) employed passive loading approaches with milk-derived exosomes to deliver paclitaxel, doxorubicin, and withaferin A. The exosome drug delivery system not only reduced the IC50 values of chemotherapeutics but also induced in vivo antitumor activity in mouse lung cancer models. Kim et al. (2016) tested paclitaxel-loaded exosomes in murine Lewis lung carcinoma cell lines. They observed that paclitaxel-loaded exosomes significantly lowered the IC50 of free drug in cancer cell lines. Results from our own research have shown that exosomes loaded with gold nanoparticles to which anticancer drugs (doxorubicin or cisplatin) were attached by a pH-stimulated linker were able to deliver drugs to H1299 and A549 lung cancer cell lines and produced therapeutic response (Srivastava, Amreddy, et al., 2016).

Thus, new avenues are opening in the field of exosome technology and can be explored for the diagnosis and treatment of lung cancer.

2. EXOSOMES

Exosomes are now the most widely studied extracellular vesicles that are produced in the cell, yet there is no classical definition for exosomes; their characteristics and definition often overlap with that of other cellular vesicles, such as microvesicles or apoptotic bodies (Akers, Gonda, Kim, Carter, & Chen, 2013; Fig. 1). To circumvent this confusion, many studies have used a combined term of extracellular vesicles (EVs) or nanovesicles (NVs); nevertheless, these studies mostly involved exosomes (Jørgensen et al., 2013; Nakano, Minata, & Rak, 2015). It is crucial to understand the basic structure and composition of exosomes in order to use them in nanomedicine.

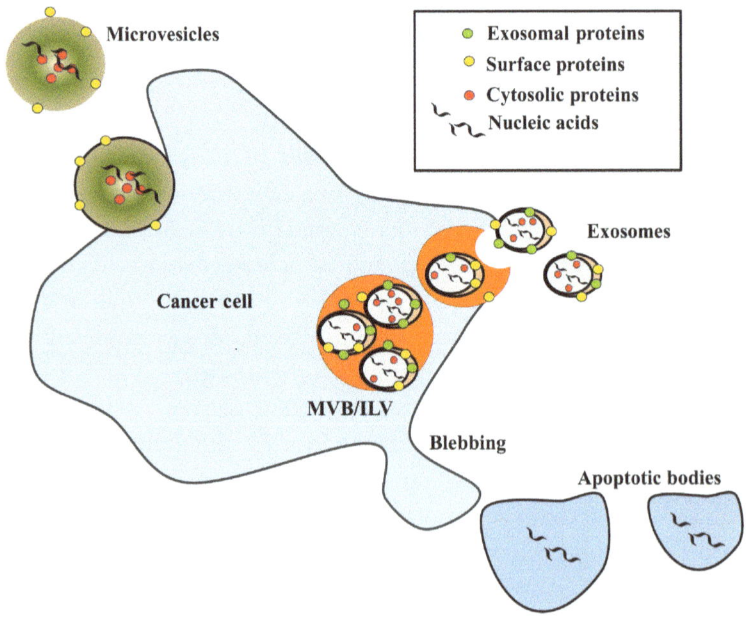

Fig. 1 Different vesicles produced by the cells.

2.1 Structure and Composition of Exosomes

The extremely small size of exosomes prevents their visualization under conventional optical microscopes. The transmission electron microscope (TEM) is the preferred method of visualizing exosomes. The isolated exosomes are directly mounted on a metallic formvar grid and are observed after negative staining with uranyl acetate. In structure, exosomes are heterogeneous in size, ranging from 30 to 150 nm in diameter. The morphology of exosomes as visualized in TEM images is largely defined as a double-membrane, cup-shaped structure (Jung & Mun, 2018; Srivastava, Amreddy, et al., 2016). However, Sharma et al. used atomic force microscopy and field-emission scanning electron microscopy to visualize exosomes, and described exosomes as having a three-dimensional, round structure. According to the study, the cup- or saucer-shaped structure of exosomes that is usually described is largely based on TEM imaging and is an artifact that is introduced due to the external force exerted on exosomes during the sample preparation (Sharma et al., 2010).

The round structure of exosomes has been confirmed using cryo-EM. Cryo-EM is considered superior to TEM, as cryo-EM does not involve dehydration or staining steps and preserves the native biological structure by vitrification, which maintains the biological specimens at atomic resolution. Exosome samples were cryo-fixed, which transforms the water present in exosomes into a glass-like state without forming ice crystals, and the samples were then visualized under cryo-EM. The exosomes appeared as rounded electron-dense structures consisting of bilayered lipoproteins (Cizmar & Yuana, 2017; Yuana et al., 2013).

Exosomes can also be characterized based on their biophysical properties. The method used to do this, however, does not illustrate the ultrastructure of exosomes, but provides information regarding the concentration and size of exosomes in the isolated samples. Two companies offer technologies for performing such analysis. The first is qNano® by Izon Science Ltd. which uses a tunable resistive pulse sensor to count the particles passing through a tunable core under the influence of electric current (Maas, Broekman, & de Vrij, 2017; Maas, De Vrij, & Broekman, 2014). The second technique is called nanotracker analysis, which is performed with a device called NanoSight (Malvern, Inc.). This device measures the particles by passing the samples under a laser light. The light scattered by exosomes is recorded by a high-resolution camera attached to an optical system. Later, using a specific computer algorithm, the track formed by each particle under the influence of light is measured, and is represented as the size and number of exosomes present (Dragovic et al., 2011; Oosthuyzen et al., 2013).

Exosomes are reported to have lipoproteins in their membrane structure, which will bear some charge. This charge can be a determining factor in designing a drug delivery vehicle. Researchers have attempted to measure the ζ (zeta)-potential of exosomes so that appropriate fabrications can be done to enable efficient cellular uptake and delivery of therapeutic payload into the recipient cells making exosomes as an amenable drug delivery vehicle. In summary, the research analyzing the characteristics of exosomes describes a size in range of 30–150 nm, a lipoprotein bilipid layer, a flat-to-round structure, and a membrane potential of −14 to −24 mV (Wang et al., 2015).

The lumens of exosomes mirror the composition of the cell from which the exosomes originate. This knowledge will be helpful in identifying the biomolecules that can be used for therapeutic purposes, as well as determining the molecules that can be loaded into them. Various researchers have identified many biologically active molecules in the lumen, including RNA, mRNA, miRNA, DNA, proteins, and lipids (Thakur et al., 2014; Théry, Zitvogel, & Amigorena, 2002; Vojtech et al., 2014). Although numerous studies have been performed to understand the composition of exosomal lumen, it is a consistent challenge to identify one class or type of molecule that is universally present in any exosome secreted by any cell type that could be regarded as a marker for exosomes. Although tetraspanins, such as CD9, CD63, and CD81, are the most common and abundant proteins present on exosomes isolated from almost any cell, no study to date has confirmed the universal presence of tetraspanins on exosomes (Srivastava et al., 2015).

Certain adhesion proteins, such as integrins and tetraspanins, are predominantly found on the exosome surface. Such proteins can be useful in using exosomes as drug delivery vehicles for appropriate receptors on cancer cells to design targeted delivery of therapeutic payloads (Batrakova & Kim, 2015).

Different cytosolic proteins, such as tubulin, actin, ANNEXINS, RAB, SNARE, and GTPase, have also been detected in exosomes originating from different cells. These proteins are involved in biogenesis, intracellular membrane fusion, and transport of exosomes (Hannafon & Ding, 2013). Other proteins of metabolic importance, such as heat shock proteins HSP70, Hsp90, and cytochrome "c," are also frequently detected in exosomes. The knowledge of exosomal content derived from different types of cells or tumor types can aid in determining and designing exosomes for therapeutic purposes. Vesiclepedia and Exocarta are online databases that meticulously curate updated information about the contents of extracellular vesicles and exosomes and can be used for such experiments

Table 1 Two Manually Curated Databases Describe the Number of Biomolecules Identified in Contents of Extracellular Vesicles (EVs) in Different Studies

	Vesiclepedia Database[a]	Exocarta Database[a]
Proteins	92,897	9769
mRNA	27,642	3408
miRNA	4934	2838
Lipids	584	1116
No. of Studies	538	286

[a]Database accessed on: March 8th, 2018.

(Table 1; Keerthikumar et al., 2015; Mathivanan, Fahner, Reid, & Simpson, 2012; Mathivanan & Simpson, 2009; Simpson, Kalra, & Mathivanan, 2012).

2.2 Biogenesis of Exosomes

Exosomes are produced in cells through a dynamic multistep endocytic pathway that is initiated by maturation of early endosomes into late endosomes. During this process, inward folding of early endosomal membranes sequesters proteins, lipids, and other biomolecules into small vesicles called internal luminal vesicles (ILV). The ILV-containing endosomes are morphologically different and are termed multivesicular bodies (MVBs). MVBs end up in lysosomes, where enzymes such as hydrolase degrade the components of MVBs, ensuring the removal of toxic materials. However, few of the MVBs escape degradation due to the presence of specific surface proteins, such as tetraspanins, CD63, and lysosomal-associated membrane proteins LAMP1 and LAMP2. The presence of these proteins enables the MVB to escape and fuse to the plasma membrane and get released to the extracellular milieu as vesicles termed exosomes (Colombo, Raposo, & Théry, 2014; Hessvik & Llorente, 2018).

In addition to biosynthesis of exosomes, whole cellular machinery consisting of about thirty proteins is involved in selective packaging of cellular cargo into exosomes during biogenesis. The proteins are members of a complex known as the endosomal sorting complex required for transport (ESCRT). The function and contribution of ESCRT in exosome biogenesis has been extensively reviewed by Colombo et al. (2013) and Stoorvogel (2015). Four protein members of ESCRT play a vital role in packaging biomolecules inside the lumens of exosomes and in the release of exosomes from the cell. Other associated or helper proteins, such as Alix and VSP4, are reported to be involved in exosome biosynthesis (Hannafon & Ding, 2013). Some ESCRT-independent mechanisms are also involved in exosome packaging and release from the cells. One such example is the ceramides, which have been shown to influence exosome biosynthesis. Blocking the production of ceramides using an inhibitor of spingomylenase, an enzyme-involved in ceramide synthesis pathway, resulted in reduced exosome production (Subra, Laulagnier, Perret, & Record, 2007; Trajkovic et al., 2008).

3. THERANOSTICS APPLICATIONS OF EXOSOMES IN LUNG CANCER

Extensive research has resulted in identification of critical molecular targets that can manipulate lung cancer cell growth by perturbing the associated signaling pathways. Researchers have also identified, isolated, and synthesized drugs and molecules to control lung cancer (Zappa & Mousa, 2016). Preclinical studies have shown promising results. In spite of these successes, the majority of such discoveries are far from human applications since there are few modalities available to deliver them efficiently and safely. Biological barriers, poor uptake by cells, and digestion by hydrolyzing enzymes or nucleases are the major reasons that many therapies cannot be used in their naïve form and must be delivered by a vehicle that can provide protection.

The structural and physiological properties of exosomes offer tremendous opportunities to utilize exosomes in lung cancer therapy and diagnosis. In 2004, Bard and colleagues carried out the first study of exosomes in lung cancer when they analyzed exosomes from pleural effusions of patients with lung cancer (Bard et al., 2004). The ability of exosomes to reach almost every part of body and to deliver their content to recipient cells is superior to that of any natural or synthetic drug carrier. The exosomal cargo can also be manipulated to package therapeutic molecules for treatment.

Due to their ubiquitous access to the body and their ability to represent their cell of origin, exosomes have been explored for diagnostic purposes. Various imaging moieties are also being used to develop sensitive and harmless imaging methods in cancers, especially in lung cancer, in which frequent tissue biopsy is not feasible. The presence of exosomes in bodily fluids has made exosomes important candidates for developing noninvasive or minimally invasive diagnostic modalities. In Sections 3.1 and 3.2, the theranostic (therapeutic and diagnostic) capabilities of exosomes are discussed in lung and other cancers.

3.1 Exosomes in Therapeutics

3.1.1 Synthetic Drug Delivery Systems

The advent of nanotechnology brought hope for overcoming the issues of sending therapeutics to cancer sites, especially in lung cancer. Nanocarriers were expected to deliver drugs to tumor cells and produce the intended cytotoxic effects while minimizing the exposure of normal cells to drugs. Liposomes are phospholipid bilayer-containing synthetic vesicles that can be used for drug delivery. With a membrane structure similar to that of cells, liposomes are biocompatible and can deliver hydrophilic (in its aqueous lumen) and hydrophobic (on its phospholipid membrane) drugs.

Liposome-based drug delivery garnered much attention in experimental therapeutics. Doxil®, a liposomal-based drug delivery vehicle for doxorubicin, has been successful in breast cancer treatment. The side effects of doxorubicin (DOx), such as cardiotoxicity, are considerably reduced in the encapsulated liposomal form (Lao et al., 2013; Perez, Domenech, Frankel, & Vogel, 2002). Due to the large surface area of the lung, a drug delivery system for lung cancer must release drug in controlled manner for a prolonged period. A liposome formulation based on dipalmitoylphosphatidylcholine is under investigation for intratracheal delivery in mice, with a view toward treating lung infections in patients with cystic fibrosis. Studies have shown better drug uptake in pulmonary cells after administration of the complex. The initial success has encouraged researchers to explore its applications in lung cancer in the future (Rudokas, Najlah, Alhnan, & Elhissi, 2016).

In addition to liposomes, many metallic nanoparticles, such as those involving gold, iron, and galladinium, have been successfully tested for drug delivery. These nanoparticles can be designed in various structures and shapes to incorporate different drugs, different amounts of therapeutics, or multiple types of agents to produce combinatorial effects. Finally, the surfaces of these metal-based nanoparticles can be easily manipulated for functionalization with targeting ligand. The biocompatibility of these nanoparticles is not yet established, but may be explored for drug delivery (Amreddy et al., 2018).

3.1.2 Exosomes: The Natural Delivery Vehicle for Therapeutic Molecules

Although various nanoparticles have been designed for drug delivery, clinical applications have failed due to poor bioavailability, nontargeted cytotoxicity, and immunogenicity (Blanco, Shen, & Ferrari, 2015; De Jong & Borm, 2008). An ideal drug delivery vehicle should overcome these limitations;

natural cellular vesicles, such as exosomes, are now being explored as drug carriers due to their lack of immunogenicity and ability to cross biological barriers (Johnsen et al., 2014). Berry-derived compounds, such as aglycones and anthocyanidins, which have antioxidant, antiproliferative, apoptotic, and antiinflammatory properties, could be useful in cancer treatment. However, due to their poor bioavailability and retention, these compounds cannot be used efficiently. In a recent study, Munagala et al. attempted to encapsulate these compounds in exosomal lumen and showed enhanced therapeutic effect in cancer cells and in lung cancer xenografts in nude mice, suggesting the benefit of using exosomes as drug delivery vehicles (Munagala et al., 2017).

One major disadvantage with currently available drug delivery systems is that they are unable to cross the blood–brain barrier. Brain endothelial cell-derived exosomes could potentially be used to deliver anticancer drug to the brain, as these exosomes can easily cross the blood–brain barrier (Yang et al., 2015). There are two major considerations regarding the use of exosomes as drug carriers: (1) loading of therapeutics and (2) the type of cargo.

3.1.3 Loading of Therapeutic Agents on Exosomes

The key requirements for using exosomes as therapeutic carriers are high encapsulation efficiency of loading molecules and stability of biomolecules and exosomal structure. The ability of exosomes to carry therapeutic molecules/genes is expedited by loading them with molecules of interest extrinsically. This task is achieved by manipulating the donor cells to incorporate the therapeutic molecules, followed by isolating exosomes primed with desired molecules. This method has been useful in loading molecules that are difficult to add to exosomes through physical methods, e.g., hydrophobic molecules that will not cross the lipid bilayer membrane. In addition, molecules such as long noncoding RNAs, miRNAs, and siRNAs can be added through genome engineering of the donor cells, which helps in de novo production and incorporation of these molecules into exosomes (Johnsen et al., 2014; Srivastava, Babu, et al., 2016).

In another method, the exosomes are first purified from the donor cells and are then loaded with desired biomolecules by in vitro methods, such as incubation, electroporation, or using chemical reagents. Since the vesicle membranes are generally dynamic in nature, small hydrophilic compounds, e.g., synthetic drugs and miRNAs, can be introduced by making reversible, transient changes in exosome membranes, and incorporating the molecules into them (Bryniarski et al., 2013; Melo et al., 2014; Zhang et al., 2010).

These methods are straightforward, but there is no control on the incorporation of molecules by the exosomes.

Curcumin is a diarylheptanoid and a tautomeric polyphenolic hydrophobic compound that possesses antiinflammatory and antioxidative properties. Its therapeutic potential has been explored. Curcumin is also the first compound that has been tested for loading on exosomes (Sun et al., 2010). Recently, Kalani et al. loaded curcumin on exosomes by both priming and physically loading on exosomes. They created CUR-EXO by treating mouse brain endothelial cells with 7.5 μM curcumin for 72 h. The exosomes were then isolated by ultracentrifugation and demonstrated enhanced therapeutic effect (Kalani, Kamat, Chaturvedi, Tyagi, & Tyagi, 2014). In another study, the exosomes were isolated from mouse embryonic stem cells and were then loaded with curcumin to create MESC-exo^{cur}, which was checked for ischemia reperfusion (IR) injury in mice. The results showed the combing the potentials of embryonic stem cells exosomes and curcumin showed effective treatment response in IR-injured mouse inking the importance of exosomes as efficient drug delivery carriers (Kalani et al., 2016).

Electroporation involves the use of weak electric pulses to make rehealable transient holes in the exosome membrane. Alvarez-Erviti et al. were among the first investigators to add exogenous siRNA into exosomes. They used electroporation to load exogenous siRNA into exosomes from dendritic cells and noted a 25%–35% siRNA loading efficiency into the lumens of exosomes. The loaded exosomes were injected back into the cells and successfully delivered the therapeutic cargo to targeted brain cells after crossing the blood–brain barrier (Alvarez-Erviti et al., 2011). In addition, siRNA for MAPK was incorporated into exosomes isolated from HeLa cells and HTB-177 lung cancer cells; peripheral blood mononuclear cells were successfully delivered into monocytes and lymphocytes, causing gene silencing and demonstrating the technique's effectiveness (Wahlgren et al., 2012). Doxorubicin was also electroporated into exosomes derived from mouse immature dendritic cells and then injected intravenously. The doxorubicin-loaded exosomes delivered the drug into targeted tumors and reduced the tumor size and growth, illustrating their potential in clinical applications (Tian et al., 2014).

In order to load therapeutic molecules, we developed a unique strategy in which we combined the synthetic nanotechnology with exosome biology. The new delivery system that we named "nanosomes" consisted of drug-conjugated gold nanoparticles loaded on and inside exosome-derived normal lung fibroblasts. This strategy is unique as it does not involve any cloning or other procedure, and the presence of gold nanoparticles permits

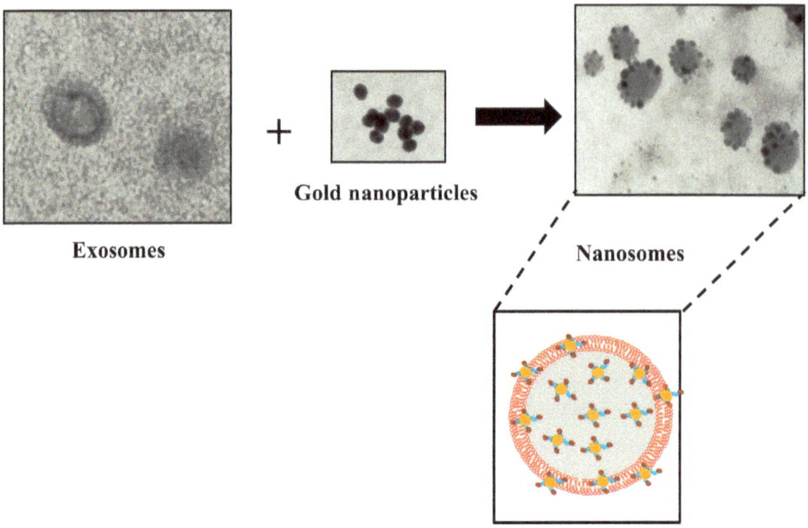

Fig. 2 Synthesis of nanosomes—an exosome-based drug delivery vehicle constructed by combining gold nanoparticles and exosomes. *Figure modified from Srivastava, A., Amreddy, N., Babu, A., Panneerselvam, J., Mehta, M., Muralidharan, R., et al. (2016). Nanosomes carrying doxorubicin exhibit potent anticancer activity against human lung cancer cells.* Scientific Reports, 6, *38541. https://doi.org/10.1038/srep38541. Copyright © 2016, Rights Managed by Nature Publishing Group.*

functionalizing the GNPs with different targeting molecules or other therapeutics (Fig. 2; Srivastava, Amreddy, et al., 2016).

Electroporation sonication is another method that involves sound waves used to disrupt the membrane, allowing the drug to enter into the lumen. Following this strategy, the hydrophobic drug paclitaxel was loaded into the exosomes derived from macrophage cells. The exosomes loaded with PTX (exo-PTX) were characterized by any changes in their size, stability, drug release, and were examined for in vitro antitumor efficacy. Interestingly, increased drug loading was observed due to reformation of the membrane as an effect of sonication. Further encapsulation of PTX into exosomes resulted in increased cytotoxicity by more than approximately 50 times in drug-resistant MDCKMDR1 (Pgp+) cells. Anticancer efficacy and colocalization of airway-delivered exosomes with cancer cells were also observed in a model of murine Lewis lung carcinoma pulmonary metastases (Kim et al., 2016). As an extension of this study, the same research group developed a new exosome-based drug delivery system intended for targeted delivery of therapeutic molecules. In their latest formulation, they added targeting moiety aminoethylanisamide-polyethylene glycol (AA-PEG) into the *exo*-PTX formulation. AA-PEG targets the sigma receptors, which are overexpressed in lung cancer cells. The results showed enhanced therapeutic outcome. The combination of targeting ability with the biocompatibility of exosome-based drug formulations offers a powerful and novel delivery platform for anticancer therapy (Kim et al., 2018).

3.1.4 Therapeutic Cargo

Exosomes are natural cellular vesicles, the lumens of which can be manipulated for loading desired drug or other molecules of therapeutic interest. Exosomes isolated from certain cell types, such as immature dendritic cells or mesenchymal stem cells, are unlikely to be immunogenic. Since the function of exosomes in cellular physiology is to carry and deliver cargo from the cell of origin to recipient cells, therapeutically loaded exosomes do not encounter resistance such as engagement with the immune system and interactions with opsonin proteins, complements, antibodies, or coagulation factors. The small size of exosomes enables them to pass through stringent biological barriers like the blood–brain barrier (Yang et al., 2015). The phospholipid bilayered membrane has many fusion and adhesion proteins that aid exosomes in directly attaching to and fusing with the recipient cells to deliver their contents. Thus, exosomes are now being explored as vehicles for efficient and specific delivery of anticancer therapeutics.

Hydrophobic drugs are otherwise difficult to deliver and are generally administered intravenously. Agrawal et al. recently attempted to administer paclitaxel orally by loading milk-derived exosomes with paclitaxel (ExoPAC). Their results showed a better reduction in tumor growth than that observed with PTX delivered intravenously in a nude mouse model of human lung cancer (Agrawal et al., 2017). Similarly, milk-derived exosomes were used to deliver curcumin. The complex was made by mixing curcumin with exosomes in the presence of 10% ethanol:acetonitrile (1:1) to form a complex called ExoCUR. Upon oral administration, the curcumin complex produced significantly higher antiinflammatory response, as measured through NF-κB activation in human lung and breast cancer cells (Aqil, Munagala, Jeyabalan, Agrawal, & Gupta, 2017).

Celastrol (CEL) is another plant-derived compound that is known to inhibit the Hsp90 and NF-κB activation pathways and has shown excellent therapeutic response in H1299 and A549 lung cancer cell lines. To explore its utility in in vivo systems, CEL was loaded into exosomes. Compared with free CEL, the CEL-loaded exosomes showed enhanced antitumor efficacy against lung cancer xenografts, without exhibiting any gross or systemic toxicity (Aqil et al., 2016). In addition, Sansone et al. recently showed packaging of the whole mitochondrial genome into the lumen of exosomes and mediated hormonal therapy-resistant breast cancer (Sansone et al., 2017). Several examples were also compiled by Johnsen et al. (2014) and Aryani and Denecke (2016), who reviewed various miRNA and siRNA loaded into exosomes for their delivery to cancer cells (Aryani & Denecke, 2016; Johnsen et al., 2014).

3.1.5 Translational Application of Exosome-Based Therapeutics: Clinical Trials

The endpoint of any research performed in experimental therapeutics is translation into clinical applications. A November 20th 2017 search of www.clinicaltrial.gov showed about 15 studies related to exosomes and various cancers, including lung cancer. Although most of the studies are related to development of exosome-based diagnostics, a few investigate the therapeutic aspects (Table 2).

In one trial, the concentration of curcumin is being evaluated for its delivery through plant-derived exosomes. Specifically, the effect of exosomally delivered curcumin on immune modulation, cellular metabolism, and the phospholipid profile of normal and malignant colon cells was evaluated in subjects who are undergoing surgery for newly diagnosed colon cancer. The effect of exosomally delivered curcumin on the production of cytokines, the changes of immune cells, and glucose metabolism by administration of 13C-glucose prior to surgical resection will also be characterized (NCT01294072). In another clinical trial, exosomes derived from grape juice are being evaluated for prevention of oral mucositis associated with chemoradiation treatment of head and neck cancer. In addition, the researchers will look for interaction between tumor exosomes and grape extract exosomes by measuring the levels of cytokines in response to tumor antigens present on tumor-derived exosomes (NCT01668849).

A dendritic cell-derived exosome (Dex)-based clinical trial is in phase II. Here, the investigators are examining the use of Dex as vaccine to impart immunotherapy in combination with metronomic cyclophosphamide (mCTX) therapy in patients with lung cancer (NCT01159288). There are also two phase I trials based on immunostimulatory exosomes for antitumor therapy. In these trials, dendritic cells of patients with stage III/IV melanoma were isolated and pulsed with MAGE3 tumor antigens. Exosomes presenting MAGE3 were isolated and readministered to melanoma

Table 2 Current Clinical Trials Based on Exosomes in Cancer

Title	Recruitment	Cancer	Interventions
Exosomes as potential prognostic and predictive biomarkers	Unknown status	Gastric cancer	
Urinary exosome for cancer diagnosis	Completed	Prostate cancer	Other: ExoIntelliScore Prostate
Exosome-mediated intercellular signaling	Recruiting	Pancreatic cancer|Benign pancreatic disease	
Vaccination trial with dendritic cell-derived exosomes	Unknown status	Nonsmall cell lung cancer	Biological: Dex2
Plant exosomes to deliver curcumin to normal and cancer tissue	Active, not recruiting	Colon cancer	Dietary supplement: curcumin|Dietary supplement: Curcumin conjugated with plant exosomes|Other: No intervention
Edible plant exosome ability to prevent oral mucositis associated with chemoradiation treatment	Recruiting	Head and neck cancer|Oral mucositis	Dietary supplement: Grape extract|Drug: Lortab, fentanyl patch, mouthwash
Tumor-derived exosomes as diagnostic and prognostic markers	Withdrawn	Breast neoplasms	
Metformin hydrochloride affecting cytokines and exosomes in cancer	Recruiting	Larynx|Lip|Oral cavity|Pharynx	Radiation: External beam radiation therapy|Drug: Metformin hydrochloride|Other: Placebo

Continued

Table 2 Current Clinical Trials Based on Exosomes in Cancer—cont'd

Title	Recruitment	Cancer	Interventions
Screening for human papillomavirus-positive oropharyngeal squamous cell carcinoma	Recruiting	Oropharyngeal cancer	
Diagnostic accuracy of circulating tumor cells (CTCs) and onco-exosome	Recruiting	Pancreatic ductal adenocarcinoma (PDAC)	Procedure: Blood samples|Procedure: Portal vein blood sample
Consistency analysis of PD-L1 in cancer tissue and plasma exosome	Not yet recruiting	NSCLC	Other: Liquid biopsy
Consistency analysis of PD-L1 in lung cancer tissue and plasma exosome before and after radiotherapy	Not yet recruiting	NSCLC	Radiation: Radiotherapy
Circulating exosome RNA in lung metastases of primary high-grade osteosarcoma	Recruiting	Lung metastases|Osteosarcoma	Other: Blood samples
Anaplastic thyroid cancer and follicular thyroid cancer-derived exosomal analysis via treatment of lovastatin and vildagliptin and Pilot Prognostic Study via urine exosomal biological markers in thyroid cancer patients	Not yet recruiting	Thyroid cancer	
ncRNAs in exosomes of cholangiocarcinoma	Recruiting	Cholangiocarcinoma|Benign biliary stricture	

patients. Therapy appeared to be well tolerated by all patients and induced the desired immune effects in some patients, showing clinical feasibility for exosome-based therapeutics (Escudier et al., 2005).

3.2 Exosomes in Diagnosis

High mortality and failure of various therapeutic interventions occurs in lung cancer because the current diagnostic approaches are not efficient enough to detect the disease at early stages. In general, lung cancer is detected when the disease has already progressed into an advanced stage. Hence, timely diagnosis of lung cancer is a crucial factor for favorable therapeutic response. In clinics, tissue collection is considered the standard practice for lung cancer diagnosis. Unfortunately, this method is insufficient to present a complete picture of disease state, especially in the era of precision and personal medicine. The limitations with tissue biopsies, such as collection of a limited amount of tissue, inability to collect tissues multiple times, and frequent tumor heterogeneity, mean that the collected localized tissue sample may not be sufficient to provide the global status of the disease. Hence, alternative routes of diagnosis are being explored.

Liquid biopsies, or analysis of bodily fluids, have drawn attention. Since exosomes can be isolated from almost all bodily fluids and their content mirrors the cell of origin, exosomes can be used to determine the state of the body. In several studies, the exosomes are manipulated in such a way that they can also be used for imaging purposes. Section 3.2 describes the various ways in which exosomes are studied for their possible role in diagnostics (Cui, Cheng, Qin, & Jiang, 2018).

3.2.1 Imaging of Exosomes

Many studies in lung and other cancers have shown that exosomes can successfully deliver the therapeutic cargo to recipient cells. Partial or preliminary success in targeted delivery of cargo to the site of cancer cells has been realized. Successful targeted delivery will lead to imaging the exosomes reaching the cancer site, which will pave the way for development of novel high-precision diagnostic and prognostic modalities. Reports of several approaches to image exosomes are available in the literature.

3.2.1.1 Labeling of Exosomes With Nanoparticles for Imaging

As an alternative to conventional fluorescent dyes, a new group of nanosized optical reporter called semiconductor quantum dots are under investigation. These quantum dots are nanosized crystals that are superior to florescent dyes, as they are more stable and have tunable optical properties that can be used for a wide range of optical applications, including in vivo imaging and diagnostics (Walling, Novak, & Shepard, 2009). Zong et al. used silicon quantum dots (Si QDs) to image exosomes using a microscopic method called single-molecule localization microscopy (SMLM). The CD63 protein present on the outer membrane of exosomes was tethered with Si QDs to form nanoprobes, which recognized exosomes. Using the SMLM, a high-resolution image of the exosome was obtained. Biocompatible in nature, Si QDs will have minimal cytotoxic effects and could be ideal for live cell imaging and diagnosis (Zong et al., 2018).

In addition, researchers have explored Gold–Carbon quantum dots (GCDs). Tumor-specific antibodies were attached to GCDs, which were attached to exosomes as described for Si QDs. These exosome-labeled nanoprobes successfully demonstrated the fluorescent imaging of exosomes and showed the intrinsic intracellular behavior of exosomes that can be used for studying the role of exosomes in cancer metastasis (Jiang et al., 2018).

In addition to quantum dots, we have demonstrated an application of exosomes as theranostics in lung cancer. We used normal lung fibroblast cell (MRC9)-derived exosomes and attached doxorubicin (Dox), an anticancer drug, and 5–10 nm super paramagnetic iron oxide Nanoparticle (SPION), creating a system we termed "fexosomes." With fexosomes, we demonstrated that we can deliver anticancer therapeutics (Dox) to lung cancer cells (H1299 and A549) and simultaneously image them using magnetic resonance imaging. This dual capacity of exosomes makes them unique for developing theranostics (Fig. 3). We further intend to exploit the gold

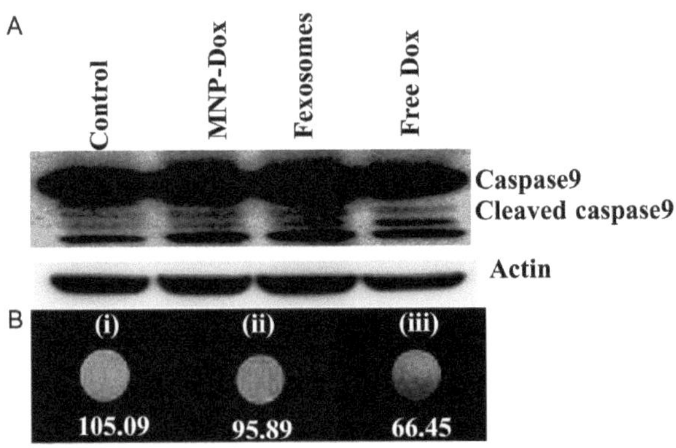

Fig. 3 Exo-MNP-Dox (fexosomes) (A) therapeutic effect was confirmed by molecular analysis of Dox-inducing DNA damage analysis through cleaved Caspase9. (B) Magnetic nanoparticles T2 MR imaging of (i) 0 μg, (ii) 12 μg, and (iii) 62 μg of MNPs in fexosomes confirm the increasing T2 relaxation.

nanoparticle part of drug carrier nanosomes described earlier in the chapter for their utility in PET imaging. Thus, the compatibility of exosomes with different nanoparticles provides an opportunity to develop novel approaches in lung and other cancer diagnosis.

3.2.1.2 Microscopic Methods for Visualizing Exosomes

Imaging exosomes can be a robust way to develop potent diagnostic and prognostic modalities. The tumor-targeted exosomes can be monitored in real time to check the distribution of exosomes and identify the precise location of tumors. Several small lipophilic fluorescent dyes, e.g., DiR, DiD, and PKH67, have been used to label the lipoprotein membranes of exosomes. Although these dyes are suitable for biodistribution studies, clinical applications for diagnosis are yet to be achieved. In addition, exosome membranes have been fluorescently labeled by tagging exosomal membrane proteins with fluorescent molecules, such as GFP or td-Tomato. This kind of labeling is considered more stable and suitable for evaluation of clinical applications.

Surface display technology is a novel approach for targeted theranostics, but we currently lack a suitable system for mammalian surface display. Stickney et al. explored exosomes and tetraspanins by adding fluorescent probes to the inner and outer surfaces of exosomes and observed the fate of exosomes in an in vivo experiment. They were able to track the exosomes in the body and cells after transfection, paving the way to visualize the fate of exosomes used for drug delivery (Stickney, Losacco, McDevitt, Zhang, & Lu, 2016). In another application of microscopic imaging of exosomes, tumor cell-derived exosomes were labeled with photo-switchable probes. This system was used to obtain super resolution imaging by using photo-activated localized microscopy or stochastic optical reconstruction microscopy. This imaging technique was successfully used to study the interaction between breast cancer cell-derived exosomes and normal cells. In the future, methods like this will be valuable for tracking the exosomes carrying therapeutic loads and will help clinicians in developing high precision and personalized treatment modalities, in addition to giving researchers a tremendous opportunity to understand the pathophysiology of cancer (Chen et al., 2018).

3.2.1.3 Spectrometric Methods of Exosome Detection

The method described thus far for detection of exosomes involves manipulation of exosomes that may have their own limitations. One such limitation is that synthesis protocols must be optimized. To avoid these synthesis-related complexities, Park et al. developed an interesting and novel approach for exosome detection. They used surface-enhanced Raman scattering spectrometry, followed by analysis with statistical pattern analysis using principal component analysis. In their study with lung cancer cell-derived exosomes, they clearly distinguished lung cancer exosomes from normal exosomes with 95.3% sensitivity and 97.3% specificity (Park et al., 2017).

3.2.2 Exosomal miRNA and Proteins as Biomarkers for Liquid Biopsy

In the current era of genomic medicine where much of the emphasis is on development of personalized and precision medicine-based standard of care, liquid biopsy is fast gaining attention. In liquid biopsy, bodily fluids (largely blood and urine) are used for diagnosis, instead of collecting tissue samples, which are generally obtained through surgical procedures. Due to its immense potential liquid biopsy-based tests are predicted to become a standard tool for diagnostic, prognostic, and predictive information in nonsmall cell lung cancer (Molina-Vila et al., 2016). Mainly circulating tumor cells (CTCs), circulating-free DNA (cfDNA), and exosomes are analyzed in liquid biopsies. Exosomes are suitable candidates for developing liquid biopsies, especially compared with CTCs and cfDNA, because the exosome remains more stable and is easily collected from urine or blood. Much of the diagnostic study related to exosomes has focused on miRNA and proteins as biomarkers. In Sections 3.2.2.1 and 3.2.2.2, these molecules are discussed with respect to their diagnostic significance in lung cancer.

3.2.2.1 Exosomal miRNA as Biomarkers

Exosomal miRNAs are considered a source of biomarkers, since they play regulatory roles and changes in physiology should be reflected in terms of changes in miRNA enrichment. Table 3 shows some of the miRNAs that could be used as biomarkers. Liu et al. reported that three miRNAs, miR-23b-3p, miR-10b-5p, and miR-21-5p, appeared to be promising prognostic biomarkers in a patient with NSCLC. They isolated exosomes from blood from 196 patients and used qPCR assay to determine the enrichment of miRNAs (Liu, Vermesh, et al., 2017; Liu, Yu, et al., 2017).

Tumor-specific circulating miRNAs have been developed as early diagnostic biomarkers for lung cancer. Remarkably, some researchers have succeeded in discovering circulating miRNAs with prognostic or predictive significance; Table 3 describes several such studies and miRNAs that have been evaluated for their diagnostic value in lung cancer. In order to substantiate the significance of miRNA in lung cancer physiology and increase confidence in exosomal miRNA-based biomarkers in NSCLC, Lai and Friedman

Table 3 List of exosomal miRNA of diagnostic significance identified in lung cancer

miRNA	Method of Study	Functional Role	References
miR-155, let-7a-2	miRNA-microarray	Differentiates lung cancer from noncancerous lung tissue	Yanaihara et al. (2006)
miR-17-3p, hsa-miR-21, hsa-miR-106a, hsa-miR-146, hsa-miR-155, hsa-miR-191, hsa-miR-192, hsa-miR-203, hsa-miR-205, hsa-miR-210, hsa-miR-212, and hsa-miR-214	miRNA-microarray	Validation of diagnostic miRNAs	Rabinowits, Gerçel-Taylor, Day, Taylor, and Kloecker (2009)
miR-29a-3p and miR-150-5p	Quantitative real-time polymerase chain reaction using Exqon miRNA expression profiling panel	Detection of response of radiotherapy in nonsmall cell lung cancer (NSCLC)	Dinh et al. (2016)
miR-21 and miR-155	miRNA-microarray	Differentiates recurrent tumors from primary tumors	Munagala, Aqil, and Gupta (2016) and Munagala, Aqil, Jeyabalan, et al. (2016)
miR-30b-5p, 30c-5p, 221-3p, 222-3p	miRNA-microarray	EGFR status	Giallombardo et al. (2016)
miR-19b-3p, miR-21-5p, miR-221-3p, miR-409-3p, miR-425-5p, and miR-584-5p	Quantitative real-time polymerase chain reaction using Exqon miRNA expression profiling panel	Biomarkers for lung adenocarcinoma detection	Cui et al. (2018)

Continued

Table 3 List of exosomal miRNA of diagnostic significance identified in lung cancer—cont'd

miRNA	Method of Study	Functional Role	References
miR-205-5p and miR-200b	Sequencing	Differentiates patients with lung cancer from patients with pneumonia and pulmonary tuberculosis	Lin et al. (2016)
miR-181-5p, miR-30a-3p, miR-30e-3p, and miR-361-5p and let-7, miR-21, miR-24, and miR-486	Seq	Identify NSCLC adenocarcinoma and squamous cell carcinoma	Jin et al. (2017)
miR-21 and miR-4257	miRNA-microarray	Predictive biomarkers for the recurrence of nonsmall cell lung cancer	Dejima, Iinuma, Kanaoka, Matsutani, and Kawamura (2017)

developed a mathematical model for early-stage NSCLC for three highest overexpressed miRs: miR-21, miR-2015, and miR-155. The model provided a simultaneous quantitative relationship between miRNA levels and tumor volumes and tumor mass. The result showed a positive correlation and is viewed as an initial step toward establishing these three miRs as a panel of biomarkers for early diagnosis of NSCLC (Lai & Friedman, 2016).

One of the challenge of using miRNA from exosome as diagnostic tool is the potential difficulty in isolating exosomes followed by purification of miRNA from them in clinical setting. In a study to obtain exosomal miRNAs for quick analysis, Wei et al. developed an electric field-induced release and measurement method. The method employs simultaneous disruption of exosomes and release of their contents for simultaneous monitoring for miRNA/protein biomarkers. This technology is currently in exploratory stages; when fully developed, this method will aid easy exosome-based diagnosis (Wei et al., 2014).

3.2.2.2 Exosomal Proteins as Biomarkers

Like miRNA, exosomes should contain proteins from the originating cell and thus should have a potential source of developing biomarkers. Profiling the whole proteome of exosomes by mass spectroscopy has revealed various biomarkers in prostate cancer and bladder cancer. A study involving 276 patients with NSCLC identified NY-ESO-1, EGFR, PLAP, EpCAM, and Alix as protein biomarkers for overall survival (Sandfeld-Paulsen, Aggerholm-Pedersen, et al., 2016). Exosomes derived from blood plasma were successfully used for evaluation of multimarker models using the Extracellular Vesicle Array (EV Array) system consisting of 37 antibodies targeting lung cancer-related proteins. Using a multimarker system the study was able to distinguish between the lung cancer and noncancer patient group. Further this method can be useful for future clinical practice as it used only 10 μL of plasma samples of performing the analysis (Jakobsen et al., 2015). In another study, using the multimarker model system Paulsen et al. found that CD151, CD171, and tetraspanin 8 were important protein markers with diagnostic value for lung cancer with different histological types. Patients with cancer of all histological subtypes vs patients without cancer showed a sensitivity and specificity index of (CD151: AUC = 0.68, $P = 0.0002$; CD171: AUC = 0.60, $P = 0.0002$; and TSPAN8: AUC = 0.60, $p = 0.0002$) (Sandfeld-Paulsen, Jakobsen, et al., 2016).

Most of the proteomic evaluation done for identification of biomarkers for lung cancer was done using blood serum. Leucine-rich α-2-glycoprotein (LRG1) was recently identified in urine-derived exosomes from patients with lung cancer and is now under investigation to explore its utility as a biomarker. Li used a nano-HPLS-chip-MS/MS method to analyze urinary exosomes and identified 18 proteins of interest that showed differential expression. The results were further validated by western blot and by immunohistochemistry of lung tissue; LRG1 was found to be expressed at higher levels in urinary exosomes and lung tissue of patients with NSCLC. These results suggest that LRG1 may be a candidate biomarker for noninvasive diagnosis of NSCLC in urine (Li, Zhang, Qiu, & Qiu, 2011).

4. TECHNIQUES FOR ISOLATION OF EXOSOMES

One bottleneck in the application of exosomes as theranostics is that exosome isolation is tedious and inefficient. At present, there is no standard protocol to ensure consistent yield of exosomes. Much initial material is needed to produce exosomes, which does not seem feasible for clinical applications. The conventional method of isolating exosomes is ultracentrifugation. This method yields a good amount of exosomes, but requires specific instrumentation and technical skill.

Recognizing this limitation, researchers are developing novel approaches to isolate exosomes from a small volume of samples. Several proprietary precipitation reagents are available from different vendors. However, these reagents are often polymer-based, which causes some impurities in the isolation. In addition, there is no defined mechanism for selective isolation of exosomes, but not microvesicles. Methods for exosome isolations using microfluidics are being developed. These methods require a small volume of input material, and isolation and sorting of specific exosomes can be performed simultaneously (Contreras-Naranjo, Wu, & Ugaz, 2017; Liang et al., 2017; Zhao, Yang, Zeng, & He, 2016). In an innovative approach, Ueda et al. constructed anti-CD9 antibody-coupled, highly porous monolithic silica microtips and used it to isolate exosomes from clinical cancer samples, including serum samples from patients with lung cancer. The isolated exosomes showed diagnostic ability, as mass spectrometry performed on the isolated exosomes identified CD91 as a lung adenocarcinoma-specific antigen on exosomes (Ueda et al., 2014).

Liu et al. recently developed a chip-based method for isolation of exosomes. Researchers reported a size-based exosome isolation tool called ExoTIC (exosome total isolation chip), which is simple, easy-to-use, modular, and facilitates high-yield and high-purity exosome isolation from biofluids. ExoTIC achieves a yield ~4- to 1000-fold higher than that with ultracentrifugation, and exosome-derived protein and microRNA levels are well-correlated between the two methods. Moreover, the researchers demonstrated that ExoTIC can sort a heterogeneous population of cancer cell exosomes based on size. They also tested ExoTIC to isolate exosomes from clinical samples, including plasma, urine, and lavage, demonstrating the device's broad applicability to cancers and other diseases (Liu, Vermesh, et al., 2017; Liu, Yu, et al., 2017).

The size-exclusion chromatography method developed by Blans et al. does not require any exosome pelleting, yet results in improved isolation of exosomes from human and bovine milk (Blans et al., 2017). Exploiting the membrane proteins expressed on the surface of exosomes, various chip-based microfluidic methods of exosome isolation have been developed. For example, Fang et al. recently developed a microfluidic chip for immunocapture and quantification of EpCAM-positive exosomes. The researchers also used a similar method to study HER2-positive exosomes from breast cancer samples (Fang et al., 2017). Improvements and innovations in isolation methods are essential for translating exosome research into clinical applications and patient care. With the development of novel methods of isolation, exosome-based theranostics will be realized in the near future.

5. CHALLENGES AND PERSPECTIVES

Research into lung cancer has identified several molecular targets that can be used for therapeutic purposes. Several tumor-targeting moieties have also been identified. Thus, lung cancer is a promising disease in which to study precision and personalized medicine. Due to their unique structure, size, composition, and mechanism of biogenesis, exosomes have a unique place in theranostics interventions (Fig. 4). The delivery of therapeutic agents through exosomes is a novel approach that holds a great future in medicine.

In diseases with complex pathophysiology and genetic makeup, like cancer, the conventional treatment regimens can become more effective if the treatment is administered through efficient, body-specific, or personalized modalities. Due to their exceptional capacity to carry and deliver biologically active molecules to different cells and due to their homing effect, exosomes are ideal candidates for developing such modalities. A number of studies show that exosomes can be loaded with anticancer therapeutic agents and can also act as therapeutic molecules by virtue of its luminal contents that contain several biomolecules that can be exploited for

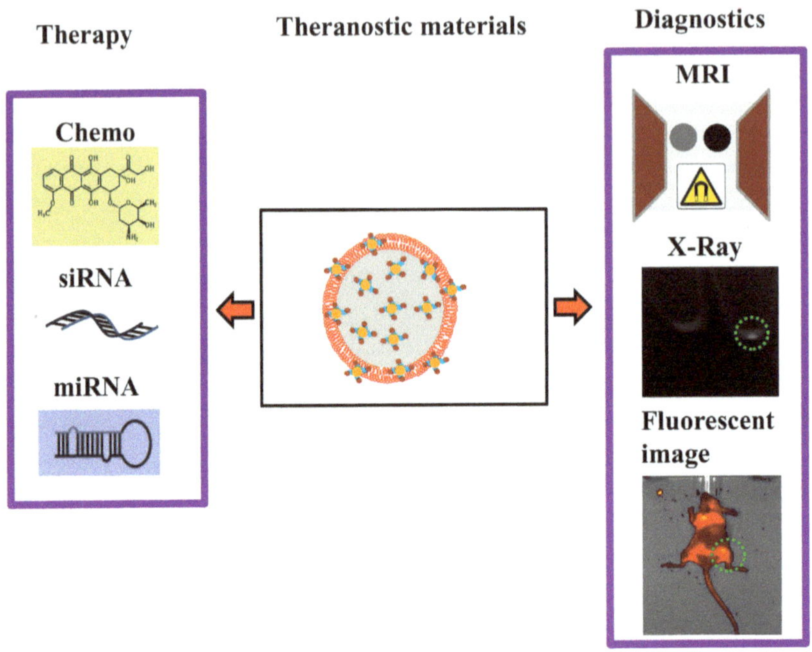

Fig. 4 The theranostic application of exosomes. The exosome can be loaded with various therapeutic molecules (*left panel*) through metallic nanoparticle. Simultaneously, the nanoparticle-loaded exosomes can be used for imaging (*top two right panels*), and the exosomes can be labeled with fluorescent dye and imaged to see the biodistribution of drugs encapsulated in fluorescent-labeled exosomes (*bottom right panel*). *Green circle* shows exosomes as imaged by X-ray and fluorescence in *right panel*.

therapeutic effect. Similarly, exosomes are promising for use in liquid biopsy, especially compared with CTCs or circulating cell-free DNAs. In summary, exosomes are unique molecules that have excellent properties to develop as theranostics.

Exosome-based theranostics are emerging as a leader in nanomedicine. However, a number of associated challenges remain, largely because of our lack of knowledge about exosomes. These challenges must be resolved before exosome-based delivery systems are applied in clinical settings.

The first and most important limitation of using exosomes in clinical settings is associated with the isolation of exosomes. Exosomes can be isolated from in vitro cell culture medium and from different bodily fluids, but the abundance of exosomes in these biofluids varies. In addition, the availability of biofluids varies. Hence, a standardized method for exosome isolation is warranted. Many methods are being investigated, but each has its own limitations. Efforts should be made to develop a standardized method that can ensure a consistent, repeatable supply of exosomes. The exosome preparation should also be of high quality, homogenous in nature, and free from other cellular vesicular contamination. Therapy sometimes involves the use of exosomes derived from specific cells, e.g., "dexosomes," or exosomes isolated from immature dendritic cells. For such studies, methods should be developed to sort dexosomes from other contaminating exosomes (Le Pecq, 2005).

In order to make use of exosomes in clinical applications for disease treatment and early diagnosis, it is essential to develop a deep understanding about the exosomes, especially their biogenesis, packaging of luminal content, and the mechanism involved in sorting, exchange, and transportation of exosomes. A number of studies that are expected to shed new light on the biology of exosomes are underway.

Finally, the foremost challenge associated with any therapeutic system of biological origin is the maintenance of good manufacturing practices (GMPs). Unlike the production of synthetic drug molecules, a slight protocol change in the production of these molecules can result in dramatic differences in the quality of final product. Hence, robust protocols to ensure the consistent production of exosomes per GMP parameters will be needed before exosome-based therapy enters clinical practice.

6. CONCLUSION

In this chapter, we highlighted the exemplary capacity of exosomes to be used as diagnostic and therapeutic agents in lung and other cancers. The properties of exosomes make them uniquely suitable for use as theranostics, and their application has been explored in many cancers. Although exosome-mediated theranostics appear promising, more research needs to be done before these theranostics are incorporated into clinical applications.

ACKNOWLEDGMENTS

The study was supported in part by a grant received from the National Institutes of Health (NIH), R01 CA167516 (R.R.), an Institutional Development Award (IDeA) from the National Institute of General Medical Sciences (P20 GM103639) of the National Institutes of Health (A.M. and R.R.), and by funds received from the Stephenson Cancer Center Seed Grant (R.R.), Presbyterian Health Foundation Seed Grant (R.R.), Presbyterian Health Foundation Bridge Grant (R.R.), and Jim and Christy Everest Endowed Chair in Cancer Developmental Therapeutics (R.R.) at the University of Oklahoma Health Sciences Center. The authors thank Ms. Kathy Kyler at the office of the Vice President of Research, OUHSC, for editorial assistance. R.R. is an Oklahoma TSET Research Scholar and holds the Jim and Christy Everest Endowed Chair in Cancer Developmental Therapeutics.

REFERENCES

Agrawal, A. K., Aqil, F., Jeyabalan, J., Spencer, W. A., Beck, J., Gachuki, B. W., et al. (2017). Milk-derived exosomes for oral delivery of paclitaxel. *Nanomedicine*, *13*(5), 1627–1636.

Akers, J. C., Gonda, D., Kim, R., Carter, B. S., & Chen, C. C. (2013). Biogenesis of extracellular vesicles (EV): Exosomes, microvesicles, retrovirus-like vesicles, and apoptotic bodies. *Journal of Neuro-Oncology*, *113*, 1–11.

Al-Nedawi, K., Meehan, B., Kerbel, R. S., Allison, A. C., & Rak, J. (2009). Endothelial expression of autocrine VEGF upon the uptake of tumor-derived microvesicles containing oncogenic EGFR. *Proceedings of the National Academy of Sciences of the United States of America*, *6*, 3794–3799.

Alvarez-Erviti, L., Seow, Y., Yin, H., Betts, C., Lakhal, S., & Wood, M. J. (2011). Delivery of siRNA to the mouse brain by systemic injection of targeted exosomes. *Nature Biotechnology*, *29*(4), 341–345.

American cancer Society. (2018). *Key statistics for lung cancer.* https://www.cancer.org/cancer/non-small-cell-lung-cancer/about/key-statistics.html. accessed on February 24th 2018.

Amreddy, N., Babu, A., Muralidharan, R., Panneerselvam, J., Srivastava, A., Ahmed, R., et al. (2018). Recent advances in nanoparticle-based cancer drug and gene delivery. *Advances in Cancer Research*, *137*, 115–170.

Aqil, F., Kausar, H., Agrawal, A. K., Jeyabalan, J., Kyakulaga, A. H., Munagala, R., et al. (2016). Exosomal formulation enhances therapeutic response of celastrol against lung cancer. *Experimental and Molecular Pathology*, *101*(1), 12–21.

Aqil, F., Munagala, R., Jeyabalan, J., Agrawal, A. K., & Gupta, R. (2017). Exosomes for the enhanced tissue bioavailability and efficacy of curcumin. *American Association of Pharmaceutical Scientists*, *19*(6), 1691–1702.

Aryani, A., & Denecke, B. (2016). Exosomes as a nanodelivery system: A key to the future of neuromedicine? *Molecular Neurobiology*, *53*, 818–834.

Bard, M. P., Hegmans, J. P., Hemmes, A., Luider, T. M., Willemsen, R., Severijnen, L. A., et al. (2004). Proteomic analysis of exosomes isolated from human malignant pleural effusions. *American Journal of Respiratory Cell and Molecular Biology*, *31*(1), 114–121.

Batrakova, E. V., & Kim, M. S. (2015). Using exosomes, naturally-equipped nanocarriers, for drug delivery. *Journal of Controlled Release: Official Journal of the Controlled Release Society*, *219*, 396–405.

Blanco, E., Shen, H., & Ferrari, M. (2015). Principles of nanoparticle design for overcoming biological barriers to drug delivery. *Nature Biotechnology*, *33*(9), 941–951.

Blans, K., Hansen, M. S., Sørensen, L. V., Hvam, M. L., Howard, K. A., Möller, A., et al. (2017). Pellet-free isolation of human and bovine milk extracellular vesicles by size-exclusion chromatography. *Journal of Extracellular Vesicles*, *6*(1), 1294340.

Bryniarski, K., Ptak, W., Jayakumar, A., Püllmann, K., Caplan, M. J., Chairoungdua, A., et al. (2013). Antigen specific antibody-coated, exosome-like nanovesicles deliver suppressor T-cell microRNA-150 to effector T cells to inhibit contact sensitivity. *The Journal of Allergy and Clinical Immunology*, *132*(1), 170–181.

Chen, C., Zong, S., Wang, Z., Lu, J., Zhu, D., Zhang, Y., et al. (2018). Gold-carbon dots for the intracellular imaging of cancer-derived exosomes. *Nanotechnology*, *29*(17), 175701.

Cizmar, P., & Yuana, Y. (2017). Detection and characterization of extracellular vesicles by transmission and cryo-transmission electron microscopy. *Methods in Molecular Biology*, *1660*, 221–232.

Colombo, M., Moita, C., van Niel, G., Kowal, J., Vigneron, J., Benaroch, P., et al. (2013). Analysis of ESCRT functions in exosome biogenesis, composition and secretion highlights the heterogeneity of extracellular vesicles. *Journal of Cell Science*, *126*, 5553–5565.

Colombo, M., Raposo, G., & Théry, C. (2014). Biogenesis, secretion, and intercellular interactions of exosomes and other extracellular vesicles. *Annual Review of Cell and Developmental Biology*, *30*(1), 255–289.

Contreras-Naranjo, J. C., Wu, H. J., & Ugaz, V. M. (2017). Microfluidics for exosome isolation and analysis: Enabling liquid biopsy for personalized medicine. *Lab on a Chip*, *17*(21), 3558–3577.

Cui, S., Cheng, Z., Qin, W., & Jiang, L. (2018). Exosomes as a liquid biopsy for lung cancer. *Lung Cancer*, *116*, 46–54.

De Jong, W. H., & Borm, P. J. (2008). Drug delivery and nanoparticles: Applications and hazards. *International Journal of Nanomedicine*, *3*(2), 133–149.

Dejima, H., Iinuma, H., Kanaoka, R., Matsutani, N., & Kawamura, M. (2017). Exosomal microRNA in plasma as a non-invasive biomarker for the recurrence of non-small cell lung cancer. *Oncology Letters*, *13*(3), 1256–1263.

Dinh, T.-K. T., Fendler, W., Fendler, J. C., Acharya, S. S., O'Leary, C., Deraska, P. V., et al. (2016). Circulating miR-29a and miR-150 correlate with delivered dose during thoracic radiation therapy for non-small cell lung cancer. *Radiation Oncology (London England)*, *11*, 61.

Dragovic, R. A., Gardiner, C., Brooks, A. S., Tannetta, D. S., Ferguson, D. J., Hole, P., et al. (2011). Sizing and phenotyping of cellular vesicles using nanoparticle tracking analysis. *Nanomedicine*, *7*(6), 780–788.

Escudier, B., Dorval, T., Chaput, N., André, F., Caby, M.-P., & Novault, S. (2005). Vaccination of metastatic melanoma patients with autologous dendritic cell (DC) derived-exosomes: Results of the first phase I clinical trial. *Journal of Translational Medicine*, *3*, 10.

Fang, S., Tian, H., Li, X., Jin, D., Li, X., Kong, J., et al. (2017). Clinical application of a microfluidic chip for immunocapture and quantification of circulating exosomes to assist breast cancer diagnosis and molecular classification. *PLoS One*, *12*(4), e0175050.

Giallombardo, M., Chacartegui, J. J., Reclusa, P., Van Meerbeeck, J. P., Alessandro, R., et al. (2016). Follow up analysis by exosomal miRNAs in EGFR mutated non-small cell lung cancer (NSCLC) patients during osimertinib (AZD9291) treatment: A potential prognostic biomarker tool. *Journal of Clinical Oncology*, *34*(15_supp), e23035.

Hannafon, B. N., & Ding, W.-Q. (2013). Intercellular communication by exosome-derived microRNAs in cancer. *International Journal of Molecular Sciences*, *14*(7), 14240–14269.

Hessvik, N. P., & Llorente, A. (2018). Current knowledge on exosome biogenesis and release. *Cellular and Molecular Life Sciences*, *75*(2), 193–208.

Jakobsen, K. R., Paulsen, B. S., Bæk, R., Varming, K., Sorensen, B. S., & Jørgensen, M. M. (2015). Exosomal proteins as potential diagnostic markers in advanced non-small cell lung carcinoma. *Journal of Extracellular Vesicles*, *4*, 26659. https://doi.org/10.3402/jev.v4.26659.

Jiang, X., Zong, S., Chen, C., Zhang, Y., Wang, Z., & Cui, Y. (2018). Gold-carbon dots for the intracellular imaging of cancer-derived exosomes. *Nanotechnology*, *29*(17), 175701.

Jin, X., Chen, Y., Chen, H., Fei, S., Chen, D., Cai, X., et al. (2017). Evaluation of tumor-derived exosomal miRNA as potential diagnostic biomarkers for early-stage non-small cell lung cancer using next-generation sequencing. *Clinical Cancer Research*, *23*(17), 5311–5319.

Johnsen, K. B., Gudbergsson, J. M., Skov, M. N., Pilgaard, L., Moos, T., & Duroux, M. (2014). A comprehensive overview of exosomes as drug delivery vehicles—Endogenous nanocarriers for targeted cancer therapy. *Biochimica et Biophysica Acta*, *1846*(1), 75–87.

Jørgensen, M., Bæk, R., Pedersen, S., Søndergaard, E. K. L., Kristensen, S. R., & Varming, K. (2013). Extracellular vesicle (EV) array: Microarray capturing of exosomes and other extracellular vesicles for multiplexed phenotyping. *Journal of Extracellular Vesicles*, *2*, 20920.

Jung, M. K., & Mun, J. Y. (2018). Sample preparation and imaging of exosomes by transmission electron microscopy. *Journal of Visualized Experiments*, *4*, 131.

Kadota, T., Yoshioka, Y., Fujita, Y., Kuwano, K., & Ochiya, T. (2017). Extracellular vesicles in lung cancer-from bench to bedside. *Seminars in Cell and Developmental Biology*, *67*, 39–47.

Kalani, A., Chaturvedi, P., Kamat, P. K., Maldonado, C., Bauer, P., Joshua, I. G., et al. (2016). Curcumin-loaded embryonic stem cell exosomes restored neurovascular unit following ischemia-reperfusion injury. *The International Journal of Biochemistry & Cell Biology*, *79*, 360–369.

Kalani, A., Kamat, P., Chaturvedi, P., Tyagi, S., & Tyagi, N. (2014). Curcumin-primed exosomes mitigate endothelial cell dysfunction during hyperhomocysteinemia. *Life Sciences*, *107*, 1–7.

Keerthikumar, S., Chisanga, D., Ariyaratne, D., Al Saffar, H., Anand, S., Zhao, K., et al. (2015). A web-based compendium of exosomal cargo. *Journal of Molecular Biology*, *428*(4), 688–692.

Kim, M. S., Haney, M. J., Zhao, Y., Mahajan, V., Deygen, I., Klyachko, N. L., et al. (2016). Development of exosome-encapsulated paclitaxel to overcome MDR in Cancer cells. *Nanomedicine*, *12*(3), 655–664.

Kim, M. S., Haney, M. J., Zhao, Y., Yuan, D., Deygen, I., Klyachko, N. L., et al. (2018). Engineering macrophage-derived exosomes for targeted paclitaxel delivery to pulmonary metastases: In vitro and in vivo evaluations. *Nanomedicine*, *14*(1), 195–204.

Lai, X., & Friedman, A. (2016). Exosomal miRs in lung cancer: A mathematical model. *PLoS One*, *11*(12), e0167706.

Lao, J., Madani, J., Puértolas, T., Álvarez, M., Hernández, A., Pazo-Cid, R., et al. (2013). Liposomal doxorubicin in the treatment of breast cancer patients. *Journal of Drug Delivery*, *2013*, 456409.

Le Pecq, J. B. (2005). Dexosomes as a therapeutic cancer vaccine: From bench to bedside. *Blood Cells, Molecules, and Diseases*, *35*(2), 129–135.

Li, X.-Q., Liu, J.-T., Fan, L.-L., Liu, Y., Cheng, L., Wang, F., et al. (2016). Exosomes derived from gefitinib-treated EGFR-mutant lung cancer cells alter cisplatin sensitivity via up-regulating autophagy. *Oncotarget*, 7, 24585–24595.

Li, Y., Zhang, Y., Qiu, F., & Qiu, Z. (2011). Proteomic identification of exosomal LRG1: A potential urinary biomarker for detecting NSCLC. *Electrophoresis*, *32*(15), 1976–1983.

Liang, L. G., Sheng, Y. F., Zhou, S., Inci, F., Li, L., et al. (2017). An integrated double-filtration microfluidic device for detection of extracellular vesicles from urine for bladder cancer diagnosis. *Methods in Molecular Biology*, *1660*, 355–364.

Lin, J., Wang, Y., Zou, Y. Q., Chen, X., Huang, B., & Liu, J. (2016). Differential miRNA expression in pleural effusions derived from extracellular vesicles of patients with lung cancer, pulmonary tuberculosis, or pneumonia. *Tumor Biology*, *37*(12), 15835–15845.

Liu, F., Vermesh, O., Mani, V., Ge, T. J., Madsen, S. J., Sabour, A., et al. (2017). The exosome total isolation chip. *ACS Nano*, *11*(11), 10712–10723.

Liu, Q., Yu, Z., Yuan, S., Xie, W., Li, C., Hu, Z., et al. (2017). Circulating exosomal microRNAs as prognostic biomarkers for non-small-cell lung cancer. *Oncotarget*, *8*(8), 13048–13058.

Luan, X., Sansanaphongpricha, K., Myers, I., Chen, H., Yuan, H., & Sun, D. (2017). Engineering exosomes as refined biological nanoplatforms for drug delivery. *Acta Pharmacologica Sinica*, *38*, 754–763.

Maas, S. L1., Broekman, M. L., & de Vrij, J. (2017). Tunable resistive pulse sensing for the characterization of extracellular vesicles. *Methods in Molecular Biology*, *1545*, 21–33.

Maas, S. L. N., De Vrij, J., & Broekman, M. L. D. (2014). Quantification and size-profiling of extracellular vesicles using tunable resistive pulse sensing. *Journal of Visualized Experiments*, (92), 51623.

Mathivanan, S., Fahner, C. J., Reid, G. E., & Simpson, R. J. (2012). ExoCarta 2012: Database of exosomal proteins, RNA and lipids. *Nucleic Acids Research*, *40*, D1241–D1244.

Mathivanan, S., & Simpson, R. J. (2009). ExoCarta: A compendium of exosomal proteins and RNA. *Proteomics*, (21), 4997–5000.

Melo, S., Sugimoto, H., O'Connell, J., Kato, N., Villanueva, A., Vidal, A., et al. (2014). Cancer exosomes perform cell-independent microRNA biogenesis and promote tumorigenesis. *Cancer Cell, 26*, 707–721.

Molina-Vila, M. A., Mayo-de-Las-Casas, C., Giménez-Capitán, A., Jordana-Ariza, N., Garzón, M., & Balada, A. (2016). Liquid biopsy in non-small cell lung cancer. *Frontiers in Medicine (Lausanne), 3*, 69.

Munagala, R., Aqil, F., & Gupta, R. C. (2016). Exosomal miRNAs as biomarkers of recurrent lung cancer. *Tumour Biology, 37*(8), 10703–10714.

Munagala, R., Aqil, F., Jeyabalan, J., Agrawal, A. K., Mudd, A. M., Kyakulaga, A. H., et al. (2017). Exosomal formulation of anthocyanidins against multiple cancer types. *Cancer Letters, 393*, 94–102.

Munagala, R., Aqil, F., Jeyabalan, J., & Gupta, R. C. (2016). Bovine milk-derived exosomes for drug delivery. *Cancer Letters, 371*, 48–61.

Nakano, G., Minata, M., & Rak, J. (2015). Extracellular vesicles in the biology of brain tumour stem cells—Implications for inter-cellular communication, therapy and biomarker development. *Seminars in Cell and Developmental Biology, 40*, 17–26.

Oosthuyzen, W., Nicole, S., Ivy, J., Turtle, E. J., Street, J. M., Pound, J., et al. (2013). Quantification of human urinary exosomes by nanoparticle tracking analysis. *The Journal of Physiology, 591*, 5833–5842.

Park, J., Hwang, M., Choi, B., Jeong, H., Jung, J. H., Kim, H. K., et al. (2017). Exosome classification by pattern analysis of surface-enhanced Raman spectroscopy data for lung cancer diagnosis. *Analytical Chemistry, 89*, 6695–6701.

Perez, A. T., Domenech, G. H., Frankel, C., & Vogel, C. L. (2002). Pegylated liposomal doxorubicin (Doxil) for metastatic breast cancer: The cancer research network, Inc. experience. *Cancer Investigation, 2*, 22–29.

Rabinowits, G., Gerçel-Taylor, C., Day, J. M., Taylor, D. D., & Kloecker, G. H. (2009). Exosomal microRNA: A diagnostic marker for lung cancer. *Clinical Lung Cancer, 10*(1), 42–46.

Rudokas, M., Najlah, M., Alhnan, M. A., & Elhissi, A. (2016). Liposome delivery systems for inhalation: A critical review highlighting formulation issues and anticancer applications. *Medical Principles and Practice, 25*, 60–72.

Sandfeld-Paulsen, B., Aggerholm-Pedersen, N., Bæk, R., Jakobsen, K. R., Meldgaard, P., Folkersen, B. H., et al. (2016). Exosomal proteins as prognostic biomarkers in non-small cell lung cancer. *Molecular Oncology, 10*(10), 1595–1602.

Sandfeld-Paulsen, B., Jakobsen, K. R., Bæk, R., Folkersen, B. H., Rasmussen, T. R., Meldgaard, P., et al. (2016). Exosomal proteins as diagnostic biomarkers in lung cancer. *Journal of Thoracic Oncology, 11*(10), 1701–1710.

Sansone, P., Savini, C., Kurelac, I., Chang, Q., Amato, L. B., Strillacci, A., et al. (2017). Packaging and transfer of mitochondrial DNA via exosomes regulate escape from dormancy in hormonal therapy-resistant breast cancer. *Proceedings of the National Academy of Sciences of the United States of America, 114*(43), E9066–E9075.

Sharma, S., Rasool, H., Palanisamy, V., Mathisen, C., Schmidt, M., Wong, D., et al. (2010). Structural-mechanical characterization of nanoparticles-exosomes in human saliva, using correlative AFM, FESEM, and force spectroscopy. *ACS Nano, 4*(4), 1921–1926.

Simpson, R. J., Kalra, H., & Mathivanan, S. (2012). ExoCarta as a resource for exosomal research. *Journal of Extracellular Vesicles, 10*, 3402.

Srivastava, A., Amreddy, N., Babu, A., Panneerselvam, J., Mehta, M., Muralidharan, R., et al. (2016). Nanosomes carrying doxorubicin exhibit potent anticancer activity against human lung cancer cells. *Scientific Reports, 6*, 38541.

Srivastava, A., Babu, A., Filant, J., Moxley, K. M., Ruskin, R., Dhanasekaran, D., et al. (2016). Exploitation of exosomes as nanocarriers for gene-, chemo-, and immune-therapy of cancer. *Journal of Biomedical Nanotechnology*, *12*(6), 1159–1173.

Srivastava, A., Filant, J., Moxley, M., Sood, A., McMeekin, S., & Ramesh, R. (2015). Exosomes: A role for naturally occurring nanovesicles in cancer growth, diagnosis and treatment. *Current Gene Therapy*, *15*, 182–192.

Stickney, Z., Losacco, J., McDevitt, S., Zhang, Z., & Lu, B. (2016). Development of exosome surface display technology in living human cells. *Biochemical and Biophysical Research Communications*, *472*, 53–59.

Stoorvogel, W. (2015). Resolving sorting mechanisms into exosomes. *Cell Research*, *25*(5), 531–532.

Subra, C., Laulagnier, K., Perret, B., & Record, M. (2007). Exosome lipidomics unravels lipid sorting at the level of multivesicular bodies. *Biochimie*, *89*, 205–212.

Sun, D., Zhuang, X., Xiang, X., Liu, Y., Zhang, S., Liu, C., et al. (2010). Novel nanoparticle drug delivery system: The anti-inflammatory activity of curcumin is enhanced when encapsulated in exosomes. *Molecular Therapy*, *18*, 1606–1614.

Thakur, B. K., Zhang, H., Becker, A., Matei, I., Huang, Y., Costa-Silva, B., et al. (2014). Double-stranded DNA in exosomes: A novel biomarker in cancer detection. *Cell Research*, *24*(6), 766–769.

Théry, C., Zitvogel, L., & Amigorena, S. (2002). Exosomes: Composition, biogenesis and function. *Nature Reviews Immunology*, *2*(8), 569–579.

Tian, Y., Li, S., Song, J., Ji, T., Zhu, M., Anderson, G., et al. (2014). A doxorubicin delivery platform using engineered natural membrane vesicle exosomes for targeted tumor therapy. *Biomaterials*, *35*, 2383–2390.

Trajkovic, K., Hsu, C., Chiantia, S., Rajendran, L., Wenzel, D., Wieland, F., et al. (2008). Ceramide triggers budding of exosome vesicles into multivesicular endosomes. *Science*, *319*, 1244–1247.

Ueda, K., Ishikawa, N., Tatsuguchi, A., Saichi, N., Fujii, R., & Nakagawa, H. (2014). Antibody-coupled monolithic silica microtips for high-throughput molecular profiling of circulating exosomes. *Scientific Reports*, *4*, 6232.

Vanni, I., Alama, A., Grossi, F., Dal Bello, M. G., & Coco, S. (2017). Exosomes: A new horizon in lung cancer. *Drug Discovery Today*, *22*(6), 927–936.

Vojtech, L., Woo, S., Hughes, S., Levy, C., Ballweber, L., Sauteraud, R. P., et al. (2014). Exosomes in human semen carry a distinctive repertoire of small non-coding RNAs with potential regulatory functions. *Nucleic Acids Research*, *42*(11), 7290–7304.

Wahlgren, J., De, L., Karlson, T., Brisslert, M., Vaziri, S. F., Telemo, E., et al. (2012). Plasma exosomes can deliver exogenous short interfering RNA to monocytes and lymphocytes. *Nucleic Acids Research*, *40*(17), e130.

Walling, M. A., Novak, J. A., & Shepard, J. R. E. (2009). Quantum dots for live cell and in vivo imaging. *International Journal of Molecular Sciences*, *10*(2), 441–491.

Wang, Y., Zhang, L., Li, Y., Chen, L., Wang, X., Guo, W., et al. (2015). Exosomes/microvesicles from induced pluripotent stem cells deliver cardioprotective miRNAs and prevent cardiomyocyte apoptosis in the ischemic myocardium. *International Journal of Cardiology*, *192*, 61–69.

Wei, F., Lin, C.-C., Joon, A., Feng, Z., Troche, G., Lira, M. E., et al. (2014). Noninvasive saliva-based *EGFR* gene mutation detection in patients with lung cancer. *American Journal of Respiratory and Critical Care Medicine*, *190*(10), 1117–1126.

Xiao, X., Yu, S., Li, S., Wu, J., Ma, R., Cao, H., et al. (2014). Exosomes: Decreased sensitivity of lung cancer A549 cells to cisplatin. *PLoS One*, *9*(2), e89534.

Yanaihara, N., Caplen, N., Bowman, E., Seike, M., Kumamoto, K., Yi, M., et al. (2006). Unique microRNA molecular profiles in lung cancer diagnosis and prognosis. *Cancer Cell*, *9*(3), 189–198.

Yang, T., Martin, P., Fogarty, B., Brown, A., Schurman, K., Phipps, P., et al. (2015). Exosome delivered anticancer drugs across the blood-brain barrier for brain cancer therapy in danio rerio. *Pharmaceutical Research, 32*(6), 2003–2014.

Yuana, Y., Roman, K., Maxim, K., Patrick, R., Koster, A., Bertina, R., et al. (2013). Cryo-electron microscopy of extracellular vesicles in fresh plasma. *Journal of Extracellular Vesicles, 2*, 21494.

Zappa, C., & Mousa, S. A. (2016). Non-small cell lung cancer: Current treatment and future advances. *Translational Lung Cancer Research, 5*(3), 288–300.

Zhang, Y., Luo, C., He, B., Zhang, J., Cheng, G., & Wu, X. (2010). Exosomes derived from IL-12-anchored renal cancer cells increase induction of specific antitumor response in vitro: A novel vaccine for renal cell carcinoma. *International Journal of Oncology, 36*, 133–140.

Zhao, Z., Yang, Y., Zeng, Y., & He, M. (2016). A microfluidic ExoSearch chip for multiplexed exosome detection towards blood-based ovarian cancer diagnosis. *Lab on a Chip, 16*(3), 489–496.

Zong, S., Zong, J., Chen, C., Jiang, X., Zhang, Y., Wang, Z., et al. (2018). Single molecule localization imaging of exosomes using blinking silicon quantum dots. *Nanotechnology, 29*(6), 065705.

FURTHER READING

Cazzoli, R., Buttitta, F., Di Nicola, M., Malatesta, S., Marchetti, A., & Pass, H. I. (2013). MicroRNAs derived from circulating exosomes as non-invasive biomarkers for screening and diagnose lung cancer. *Journal of Thoracic Oncology: Official Publication of the International Association for the Study of Lung Cancer, 8*(9), 1156–1162.

CHAPTER TWO

Nanotechnology Approaches to Improving Cancer Immunotherapy

C. Tilden Hagan IV*,†,‡, Yusra B. Medik*,†, Andrew Z. Wang*,†,1

*Laboratory of Nano- and Translational Medicine, Lineberger Comprehensive Cancer Center, Carolina Center for Cancer Nanotechnology Excellence, Carolina Institute of Nanomedicine, University of North Carolina at Chapel Hill, Chapel Hill, NC, United States

†Department of Radiation Oncology, Lineberger Comprehensive Cancer Center, University of North Carolina at Chapel Hill, Chapel Hill, NC, United States

‡UNC/NCSU Joint Department of Biomedical Engineering, University of North Carolina at Chapel Hill, Chapel Hill, NC, United States

[1]Corresponding author: e-mail address: zawang@med.unc.edu

Contents

Advances in Cancer Research, Volume 139
ISSN 0065-230X
https://doi.org/10.1016/bs.acr.2018.05.003

Abstract

Cancer immunotherapy is a powerful, growing treatment approach to cancer that can be combined with chemotherapy, radiotherapy, and oncosurgery. Modulating the immune system to enhance anticancer response by several strategies has yielded improved cancer survival. Despite this progress, the success rate for immunotherapy has been below expectations due to unpredictable efficacy and off-target side effects from systemic dosing. Nanotechnology offers numerous different materials and targeting properties to overcome many of these challenges in immunotherapy. In this chapter, we review current immunotherapy and its challenges as well as the latest nanotechnology applications in cancer immunotherapy.

1. IMMUNOEDITING: IMMUNE RESPONSE TO TUMOR GROWTH

In the process of "immunoediting" the immune system shapes tumor progression in two opposite ways: by protecting the host against cancer and by conditioning tumor immunogenicity through three phases: elimination, equilibrium, and escape (Chen & Mellman, 2013). In the elimination phase, tumor cells activate the innate immune system due to mechanical disruption of healthy tissues. IFN-gamma secretions from macrophages and dendritic cells (DCs) cause direct cell death and induce angiostatic chemokine expression in tumor cells. Shedded antigens and other proteins of dying cancer cells are taken up by antigen-presenting cells (APCs) such as dendritic cells and macrophages and delivered to draining lymph nodes (DLNs) to activate CD4+ T cells and in turn CD8+ T cells. Tumor-associated antigen (TAA)-activated T cells circulate to the tumor microenvironment and induce cell death by releasing cytotoxic proteins such as perforin or granzyme, or by the Fas/FasL pathway. The IFN-gamma-induced proinflammatory tumor microenvironment enhances apoptosis through a feedback loop of production of more IFN-gamma and IL-12. Genetically mutated immune-resistant cancer cells that survive the elimination phase transition to equilibrium. At this point, tumor cells go into a quiescent phase and there is a balance between immune attack and cancer cell proliferation that prevents tumor outgrowth (Dunn, Bruce, Ikeda, Old, & Schreiber, 2002).

As the immune system inhibits proliferation of immune sensitive variants, remaining resistant cancer cells acquire mutations that protect them from immune detection and create an immunosuppressive tumor microenvironment. Two well-known escape mechanisms are decreasing TAA presentation and loss of MHC1 expression which binds to T cell receptors (TCRs) on CD8+ T cells. Also, cancer cells produce suppressive cytokines such as TGF-beta and IL-10 as well as inhibitory costimulatory molecules such as CTLA-4, PD-1, and PD-L1. Regulatory T cells, myeloid-derived dendritic cells, and tumor-associated macrophages contribute to tumor survival and expansion (Schreiber, Old, & Smyth, 2011).

2. CURRENT IMMUNOTHERAPIES

Cancer Immunotherapy was first investigated over one hundred years ago when Dr. Coley put forward the idea of inducing the immune system to create an antitumor response by injecting "Coley's Toxins" into resected sarcoma sites (Coley, 1898). Since then many efforts have been devoted to understanding the immune system's role in cancer progression. Rosenberg showed sustainable cancer regression with the use of high dose bolus IL-2 injection for metastatic melanoma and thus received approval from the FDA as the first immunotherapy for human cancer (Rosenberg, 2014). Checkpoint inhibitors (CPIs) were the next significant immuotherapy advancement. CTLA-4 antibody ipilimumab (Yervoy, Bristol-Myers Squibb) has been shown to improve survival and was approved by the FDA for metastatic melanoma in 2011 with PD-1/PD-L1 antibodies to follow. Clinical trials for use of CPIs on many solid and hematological cancers are in progress (Pardoll, 2012) (Fig. 1).

2.1 Checkpoint Inhibitors

Immune CPIs have advanced cancer immunotherapy and been approved as a part of the treatment model for a number of cancers including advanced melanoma (Patel & Minn, 2018).

CTLA-4 is expressed on T cells and plays role in early inhibition of T cell activation after priming in lymph nodes. It is a homologue of CD28, and strongly attaches to CD80 and CD86 on APCs and down-regulates T cell activation and increases Treg activity. CTLA-4 inhibitor Ipilimumab (Yervoy, Bristol-Myers Squibb) was the first monoclonal antibody (mAb) that was approved by the FDA in 2011 for advanced melanoma, and it is undergoing clinical trials for other solid tumors (Topalian, Drake, & Pardoll, 2012).

The PD-1: PD-L1 pathway regulates the dynamic balance between potentially pathogenic self-reactive T cells and Tregs by inhibiting the

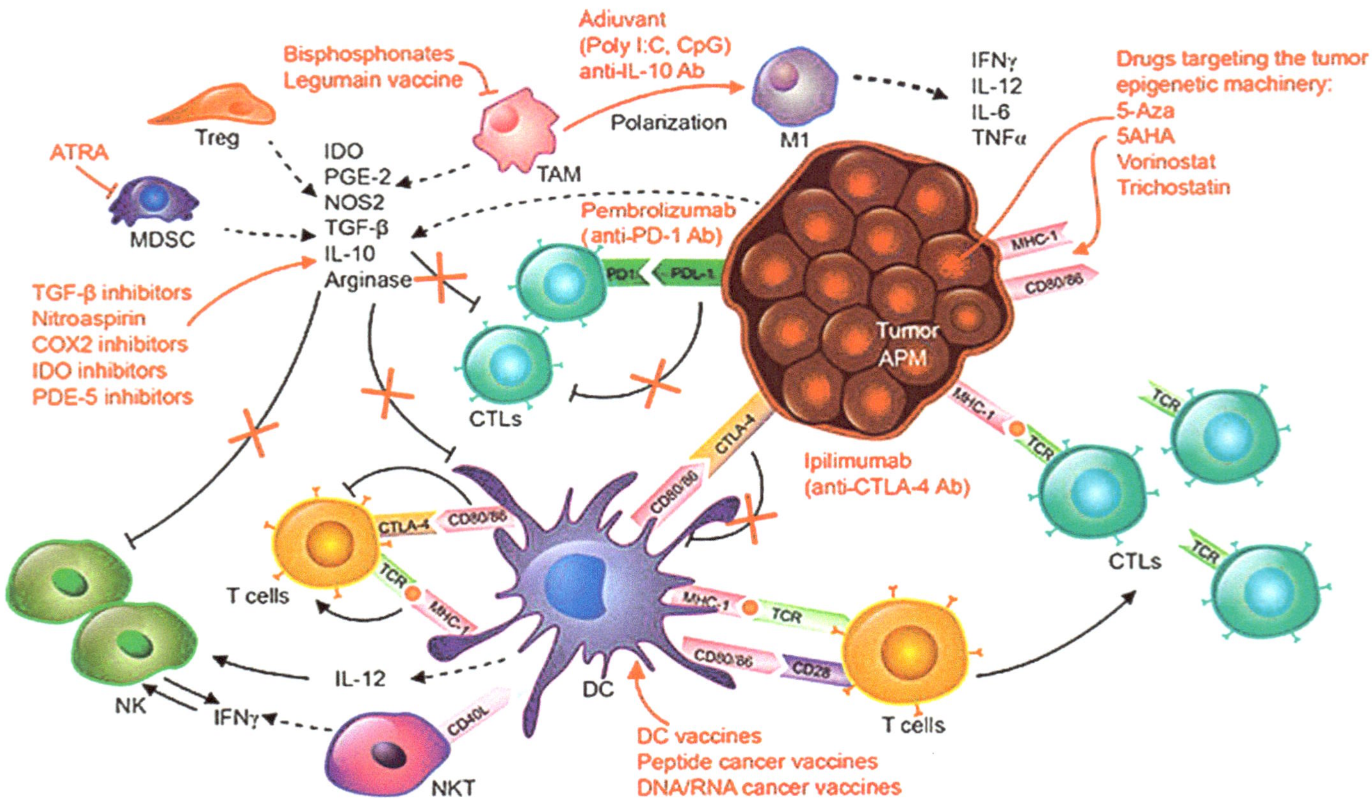

Fig. 1 Immunosuppressive tumor microenvironment and immunotherapies.

T cells and promoting the development and function of regulatory T cells. The PD-1 inhibitor Pembrolizumab was approved by the FDA in 2015, as a first-line treatment of non-BRAF mutated melanoma (Topalian et al., 2012).

CPIs may however cause immune-related adverse effects in the skin, gastrointestinal tract, endocrine tissues and liver (Boutros et al., 2016).

2.2 Adoptive Cell Transfer

Ex vivo manipulation and activation of T cells to induce a robust immune response have been researched through three different approaches: (1) isolating tumor infiltrating lymphocytes (TILs), (2) synthetic TCR gene transfer, and (3) chimeric antigen receptor expressing T cells (CAR-T).

CAR-T cells are externally manipulated T cells that express a peptide–HLA complex-independent antigen receptor (a small chain FC), a transmembrane protein and a TCR intracellular domain which is combined with costimulatory molecules such as 4-IBB and CD28 (Xie, Wei, & Tang, 2018). The optimum target for CAR-T should either be exclusively expressed on cancer cells or on nonvital tissues. CD19 is a surface antigen expressed on B lymphocytes and is an ideal target since it is unique to B cells. CD19 targeted CAR-T cells are commonly tested and shown to be effective in large clinical trials. Tisagenlecleucel (Kymriah™), a CD19 targeted autologous T cell containing the 41BB costimulatory signal, exhibited an 83% overall remission rate in relapsed B-ALL patients and was approved by the FDA to treat children and young adults with B-ALL in 2017. This approach is less effective in solid tumors and requires extensive labor and skills. There are also possible life threatening side effects which may be caused by CAR-T and TCR engineered cells such as cytokine release syndrome and severe autoimmune reactions (Khalil, Smith, Brentjens, & Wolchok, 2016).

2.3 Cancer Vaccines

Various cancer vaccine strategies have been tested including cancer peptides/proteins, killed whole tumor cells, pulsed dendritic cells, and viral proteins. Most cancer vaccines consist of tumor-specific antigens and adjuvants to target and enhance tumor-specific T cells and to cause existing T cells to proliferate.

Identification of TAAs with DNA and RNA sequencing facilitates peptide vaccine generation. They have so far been ineffective except for the Human Papilloma Virus (HPV) subtype vaccine which is a virus-protein vaccine that prevents cervical neoplasias.

Dendritic cells are professional APCs which strongly induce T cells. Dendritic cell-based vaccine Sipuleucel-T (Provenge®) was developed by isolation of autologous peripheral blood mononuclear cells and activation with PAP–GM-CSF, and was approved for hormone refractory prostate cancer.

2.4 Monoclonal Antibodies

Monoclonal antibodies (mAbs) are designed for different strategies such as labeling cancer cells to make them more susceptible to chemotherapy or to act as a targeting agent for cancer cell-specific chemotherapy. Rituximab, a CD20 antibody, was the first mAb approved by the FDA for B cell non-Hodgkin Lymphoma.

Another group of commonly used mAbs are growth factor antibodies to block oncogenic pathways. Her2, a tyrosine kinase receptor which is highly expressed on breast cancer cells, has been targeted by the antibody Transtuzumab (Herceptin®), which was approved by the FDA in 1998 for aggressive breast cancers (Chiavenna, Jaworski, & Vendrell, 2017).

In spite of these advances in understanding cancer's immune evading mechanisms and new therapy modalities, the success rate for cancer immunotherapy has been less than desired. There are several reasons for this including the immune-resistant tumor microenvironment, dosing and efficacy problems, and the extensive labor and substantial costs of possible treatments (van der Burg, Arens, Ossendorp, van Hall, & Melief, 2016).

3. NANOPARTICLE APPROACHES TO ENHANCE CANCER IMMUNOTHERAPY

Nanoparticle systems have been expanding in use in medicine and have many advantages over conventional methods. Nanoparticles can act as a protective delivery vehicle for many types of cargo, increase their cargo's stability and solubility, prolong its half life, and can be manipulated to target cancer cells (Qiu, Min, Rodgers, Zhang, & Wang, 2017).

There are many different applications for which NPs can be used for immunotherapy in cancer. These include the delivery of antigens and adjuvants as vaccines, the delivery of antibodies, targeting specific cells such as APCs or dendritic cells even more specifically, and modifying the tumor microenvironment to counteract many of the immunoinhibitory effects of tumors.

3.1 Antigen and Adjuvant Delivery

It has been shown that delivery of TAAs can cause APCs to mature which in turn activates tumor-specific T cells in lymph nodes. APCs such as dendritic cells and macrophages can uptake exogenous antigens and cross-present them in the context of Major Histocompatibility Complex-1 which is essential to develop cytotoxic CD8+ T cells. Cancer vaccines consisting of TAAs and/or adjuvants prime APCs to express TAAs and induce an immune response. Administering cancer vaccines in nanocarriers provide sustained release, stability, and enhanced immunogenicity to their cargo. Encapsulation of tumor-associated and model antigens (including Trp2 and OVA) in combination with adjuvants such as toll-like receptor agonists within varied nanomaterials has been investigated. Lipid–calcium–phosphate (LCP) nanoparticles have been fabricated to encapsulate Trp2, a melanoma-specific antigen, with CpG adjuvant in a calcium core and have been functionalized with mannose to target dendritic cells. This design resulted in efficient dendritic cell uptake, sustained antigen release, and inhibition of tumor growth in a melanoma model (Vasievich, Ramishetti, Zhang, & Huang, 2012; Zhuang et al., 2016). Poly(lactic-*co*-glycolic acid) (PLGA), liposomes, and cationic lipid DOTAP have been proven to deliver Trp2 antigens. PLGA polymers show slow release rates and prolonged immune responses compared to liposomes which have a phospholipid bilayer on the surface (Demento et al., 2012). Liang et al. activated dendritic cells and induced antitumor CD8+ T cells through liposome-coated gold nanoparticles targeting dendritic cell receptor aCD11c, as well as delivering adjuvant monophosphoryl lipid A (MPLA) and Trp2 antigen. This two-layered particle design enables protection of Trp2 from degradation and conjugation of anti-CD112 antibody on the surface (Liang et al., 2017). Kuai et al. engineered synthetic high-density lipoprotein nanodiscs which are 10 nm in size and are able to conjugate multiple personal neoantigens identified via DNA exome sequencing and adjuvants by simple mixing. It has been shown that nanodiscs carriers generate prolonged T cell responses against tumor-specific antigens and delayed tumor growth in B16F10 tumors (Kuai, Ochyl, Bahjat, Schwendeman, & Moon, 2017).

Hassan et al. showed that integrated OVA antigen onto needle-like nanoscopic carriers, carbon nanotubes (CNTs), along with CpG and anti-CD40 induced OVA-specific cellular and humoral immune responses as well as delayed tumor growth in mouse melanoma models. Compared to PLGA encapsulation whose cellular uptake relies on energy-dependent mechanisms, CNTs can enter cells through both energy-dependent and passive routes which results in higher antigen delivery. Further, researchers examined the difference between loading CNTs with a mix of free OVA and CpG ((OVA)S −/+(CpG)) vs loading CNT with a chemical conjugate of CpG and OVA(S −/+(OVA − CpG)). Codelivery of OVA and CpG in the conjugate form produced higher anti-OVA IgG and IgG2c antibodies (Hassan et al., 2016).

Size, design, and structure of nanoparticles affect the immune response, pace of trafficking to DLNs, and uptake by APCs. Duan et al. utilized pH-responsive lanthanide ions and a guanine monophosphate delivery system which releases antigens and CpG upon degradation of the metal–organic framework in the acidic environment of cytosol and endosomes which subsequently increase cross-presentation to CD8+ T cells in melanoma models (Duan et al., 2017).

Cancer cell membrane-coated cancer vaccines have been designed to overcome the challenges of peptide-based vaccines arising from the high mutation rate of cancer cells and variability in antigen expression. Kang et al. coated PLGA particles with B16F10 cell membrane to present the whole array of membrane proteins to dendritic cells which leads to increased IFN-γ production in vitro. They engineered artificial, highly immunogenic cancer cell like NPs, consisting of a CpG loaded calcium core, alongside B16OVA cell membrane and α-helix SP70 protein on an outer phospholipid bilayer. Results showed augmentation of dendritic cells and sustained antigen presentation in lymph nodes in mouse melanoma models (Kang et al., 2018).

Nanoparticles can act not only as carriers, but can also directly interact with immune cells and serve as immunostimulatory agents. Virus-like proteins are noninfectious, stable, and adaptable carriers, and some virus-like particles (VLPs) exhibit immunogenic properties. Inhalation of empty Cowpea mosaic virus (CPMV) led to a decreased tumor burden in a melanoma lung metastasis model mediated by the increased immune response, activated neutrophils, and primed CD4+ T cells (Lizotte et al., 2016).

3.2 Targeting Antigen-Presenting Cells

Dendritic cells are professional APCs and are potent inducers of T and B cell responses. Several DC cell membrane proteins have been targeted including mannose, fucose *N*-acetyl glucosamine, anti-CD11c, and anti-DEC205. Antigen-loaded PLGA cores functionalized with DC antibodies on a polyethylene glycol (PEG)-lipid layer exhibited facilitated uptake by

DCs (Cruz et al., 2014). Bocanegra Gondan and her colleagues engineered zinc doped, radiolabeled, magnetic nanoparticle filled micellar systems carrying a combination of two TLR agonists polyIC and R837 that synergistically activate macrophages and DCs. When compared to free polyIC-R837, nanoparticle delivery provides sustainable stimulation of immune cells and increased secretion of proinflammatory cytokines. Also, researchers showed that when coadministered with mSPION-OVA antigen, mZnSPION-polyIC-R837 yields a better adoptive immune response. Due to Gallium-67 content, it is possible to monitor these nanoparticle systems via MRI and SPECT imaging (Bocanegra Gondan et al., 2018).

Genetic editing of APCs to promote their maturation and activation has also been tested. Lipid nanoparticles are suitable agents for RNA delivery. Ionizable cationic lipids incorporated with PEG carrying PD-L siRNA knocked out PD-L1 and PD-L2 on dendritic cells and provoked a CD8+ T cell response without limiting migratory functions (Hobo et al., 2013). Encapsulation of a miR-148a inhibitor in polypeptide micelles along with TLR agonists prevents the TLR recognition inhibition normally caused by miR-148a in dendritic cells and results in tumor-associated dendritic cell maturation and inhibition of the immunosuppressive tumor microenvironment (Liu et al., 2016). Ray et al. used arginine-coated gold nanoparticles to deliver a CRISPR–Cas9 nuclease system knocking out SIRP-α gene in macrophages which resulted in enhancement of phagocytosis of cancer cells (Ray et al., 2018). Some nanoparticles alone modulate immunity of APCs. Polyhydroxylated fullerenols are shown to activate macrophages and prevent cancer metastasis when those activated macrophages were transfused back into a lung metastasis model. Interestingly, suppressive effects of fullerenols are correlative with electronegativity level of the particle (Tang et al., 2016).

3.3 Targeted Antibody Delivery

There are a growing number of mAb therapies approved by the FDA for varying indications such as infectious disease, immune modulation for transplant rejection or autoimmune control, and cancer. These mAbs have highly specific binding abilities which decrease off-target effects and improve cancer therapy, but also have many limitations such as poor pharmacokinetics, tissue accessibility, limited tumor penetration, impaired immune system interactions, and difficulty in crossing biological barriers (Colzani et al., 2018; Qiu et al., 2017). There are also possible risks, with serious possible side effects such as cardiotoxicity, cytokine release syndrome, infection, and autoimmune disease (Hansel, Kropshofer, Singer, Mitchell, & George, 2010). To further limit these side effects and direct delivery of mAb to the tumor microenvironment, these mAbs can be encapsulated within NPs. Colzani et al. encapsulated the HER2 targeting antibody trastuzumab (TZ, an FDA approved treatment for breast cancer), within PLGA NPs to aid in delivery and allow a time release formulation, and to also couple TZ with doxorubicin in a dual loaded NP for combined chemoimmunotherapy. They were able to achieve a 37.5% encapsulation efficiency and 2.9% loading of TZ with 78% of encapsulated mAb released over 30 days. In vitro studies showed that HER2-positive cells treated with TZ loaded NPs for 48 h offered a 92.2% reduction in binding of fluorescently labeled TZ post treatment, demonstrating the effective encapsulation and release of TZ from PLGA NPs. These loaded NPs also enabled improved degradation of HER2 over TZ alone by continuing to the lysosomal pathway unlike unencapsulated TZ. Another recent study by Mi et al. did not encapsulate the antibodies but instead conjugated anti-PD-1 and anti-OX40 antibodies to maleimide-terminated PEG-PLGA NPs (Mi et al., 2018). By using αPD-1 to block T-cell inhibition and αOX40 to increase T-cell activation, they were able to significantly upregulate T-cell activation by controlled spatiotemporal codelivery of conjugated antibodies, compared to dual administration of free antibodies. When compared in vivo, the dual antibody NPs resulted in a 30% cure rate of C57BL/6 mice, with 5 of 6 cured mice also maintaining resistance upon tumor rechallenge. These mice also demonstrated an increased survival time by more than 20% compared to any other treatment, and a significantly higher percentage of CD8+ tumor infiltrating T cells (85%) compared to the free antibodies (69%).

3.4 Gene Delivery

The delivery of RNA offers a method for intracellular antigen synthesis; however, systemic injection results in problems with inefficient targeting and uptake, and degradation by ribonucleases (Kranz et al., 2016; Sahin, Karikó, & Türeci, 2014; Yin et al., 2014). Kranz et al. developed a lipid/RNA lipoplex to target RNA delivery to DCs, increase uptake efficiency, and protect the RNA from ribonucleases. To target DCs, overall particle charge was varied by modifying lipid:RNA ratios, with particles with an excess negative charge being highly selective for spleen, lymph nodes, and bone marrow. Mice injected with lipoplexes including RNA coding for hemagglutinin showed an increase in CD4+ and CD8+ T cells with the HA-specific T-cell receptor. Another lipoplex variant with ovalbumin coding RNA was able to fully immunize mice after three rounds of treatment with no mice growing tumors within 20 days of challenge of B16-OVA, but all mice treated with RNA alone were tumor-loaded. This NP has even moved on to a phase I dose-escalation clinical trial, with RNA encoding four tumor antigens (NY-ESO-1, MAGE-A3, tyrosinase, and TPTE) included in the lipoplex. Three patients had been enrolled and received subtherapeutic dosing with all three tolerating treatment and showing only temporary flu-like symptoms. One patient showed regression of a suspected metastatic lymph node after treatment, another receiving treatment following tumor resection and remaining tumor free (7 months at time of report), and the third with eight lung metastases remaining stable after treatment. Oberli et al. similarly used a lipid NP to deliver RNA to DCs, macrophages, and neutrophils to induce a CD8+ T-cell response (Oberli et al., 2017). They however chose tyrosinase-related protein 2 (TRP2) and a mutated glycoprotein 100 (gp100) as the RNA coded antigens within the NPs, and a B16F10 melanoma tumor model. Particles from 50 to 150 nm were tested with no correlation between size and T-cell activation observed; however, the high T-cell activations were observed with particles having zeta potentials between −15 and −3 mV. All groups treated with RNA encapsulated NPs outperformed control groups, whether treated with RNA coding for only one of the chosen antigens or both. The group treated with RNA coding for both antigens was the only group with a single mouse to survive through the end of the 60-day study with no visible tumor development.

3.5 Cytokine Delivery

Cytokines play an important part in immune cell signaling and are able to direct the immune response. Correct delivery of free cytokines is prohibitive in a systemic administration due to degredation, excretion, nonspecific binding, and toxic side effects (Christian & Hunter, 2012). Most cytokines act in a local capacity, and to obtain the desired immune effects a systemically toxic concentration would be needed, with maximum safe dosing yielding subtherapeutic results. By encapsulating cytokines in targeted NPs, sufficient cytokine concentration can be delivered to illicit the immunotherapeutic response desired. Bauleth-Ramos et al. have recently developed a NP to deliver combination chemoimmunotherapy by loading the chemotherapeutic

Nutlin-3a (Nut3a) and the inflammatory cytokine granulyte-macrophage colony-stimulating factor (GM-CSF) into an acetalated dextran (Bauleth-Ramos et al., 2017). The cytokine GM-CSF is able to increase leukocyte activation and proliferation, and recruit APCs, thereby stimulating the systemic immune response (Hercus et al., 2009; Yan, Shen, Tien, Chen, & Liu, 2017). They were able to target tumor cells through the use of acetelated dextran which is a pH-sensitive polymer and will degrade faster in an acidic environment such as a lysosomal compartment or tumor tissue. The NPs showed high safety profiles up to 250 μg/mL in peripheral blood mononuclear cells, with cell viability at 180% after 72 h at 100 μg/mL, yet MCF-7 breast cancer cells showed only approximately 60% viability at the same time and concentration. There was also a significant increase in expression of the cell surface markers CD83 and CD86 on DCs and lymphocyte proliferation. Bauleth showed the coloaded NPs were able to prime naïve T cells and cause their proliferation, stimulate APCs, and alter the cytokine environment. Another recent work by Guimarães et al. discussed possible NP optimization strategies for delivery of the cytokine tumor necrosis factor-related apoptosis-inducing ligand (TRAIL) which binds to death receptors 4 and 5 to trigger apoptosis (Guimarães et al., 2018). TRAIL has been shown to induce cancer apoptosis with limited toxicity in preclinical trials and has also shown low toxicity in initial clinical trials. It has not, however, had any significant therapeutic efficacy, due to low potency, a short half-life, and poor delivery. De Miguel et al. demonstrated one possible such NP delivery vehicle by coating a lipid-based NP with TRAIL (De Miguel et al., 2016). In their NSCLC cell model they found that NP formulations significantly improved the bioactivity of TRAIL compared to the soluble form. When given in combination with a CDK inhibitor as a cell sensitizer, the TRAIL-coated NP was especially effective with cell viability reduced to 9.7% after treatment compared to 28.9% viability for the soluble combination with both at 1 μg/mL.

4. NANOPARTICLE'S STRUCTURES FOR CANCER IMMUNOTHERAPY

Nanoparticles can deliver bioactive molecules to increase therapeutic efficacy and reduce side effects through improved pharmacokinetics and biodistribution of their cargo. There are a number of nanoparticle properties and routes of administration which can be utilized to tune the particle for a particular purpose. Optimal pharmacokinetics and biodistribution typically involve particles of approximately 100 nm, a surface Zeta-potential between $\pm$ 10 mV, and with a PEG surface layer. NPs are not only useful for drug delivery, but also in directly targeting the immune system to produce cytokines and induce humoral and cellular immunity. A number of different types of nanoparticles exist including polymers, lipids, dendrimers, VLPs, and inorganics. The most common immunotherapy applications of NPs are to deliver specific antigens and adjuvants to increase uptake via APCs, most commonly DCs, in order to upregulate B and T cell activation. While TAAs are usually those introduced via NPs, they often do not illicit enough of an immunogenic response to activate APCs due to similarities to self, but by including adjuvants this response can be increased (Silva, Videira, Gaspar, Préat, & Florindo, 2013). Some NPs even act as adjuvants themselves, stimulating B- and T-cell responses, without the addition of exogenous adjuvants (Dwivedi, Tripathi, Ansari, Shanker, & Das, 2011; Zolnik, González-Fernández, Sadrieh, & Dobrovolskaia, 2010) (Fig. 2).

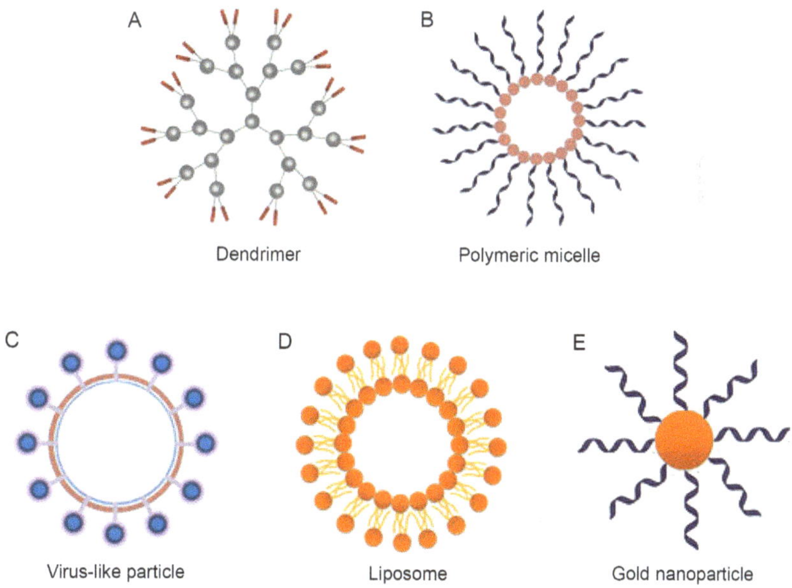

Fig. 2 Nanoparticles for cancer immunotherapy. Nanoparticles with different structures can be used cancer immunotherapy, which includes (A) dendrimer, (B) polymeric micelles, (C) virus-like nanoparticles, (D) liposomes, (E) gold nanoparticles, etc.

4.1 PLGA

PLGA is a popular polymer used for nanoparticles. It is common because it is an FDA approved, biocompatible, and biodegradable polymer, which can be used to encapsulate many biologically active compounds. These NPs have tunable characteristics such as size, morphology, surface functionalization, and high loading of antigens, adjuvants, and drugs (Demento et al., 2012; Irvine, Hanson, Rakhra, & Tokatlian, 2015). PLGA NPs also have limitations which include possible cytotoxic degradation products, and loss or inactivation of payload during preparation and prior to treatment. Polymeric micelles can also be formed from PLGA and other polymers when a hydrophilic head such as PEG is added to one end of the polymer, creating an amphiphilic copolymer. PEGylated polymers can be useful in increasing NP circulation time by preventing protein absorption and preventing interaction with the reticuloendothelial system (Grimaldi, Incoronato, Salvatore, & Soricelli, 2017). Kim utilized PLGA NPs with an encapsulated TLR7/8 bispecific imidazoquinoline-based ester agonist to enhance cytokine secretion when compared to a TLR7 monoselective agonist (Kim et al., 2018). These NPs had the most significant effect when administered subcutaneously where they then traveled to the DLNs and triggered DC activation and expansion. This led to an increased cytotoxic T lymphocyte (CTL) response through antigen-specific CD8+ T cells. They showed prophylactic and therapeutic efficacy and a reduction in systemic metastasis in numerous cancer models including renal, bladder, and melanoma cancer models. While Kim's PLGA NP only included an adjuvant, Jahan utilized a combination of encapsulated ovalbumin antigen, MPLA adjuvant, and a surface modifying Ab ligand which improved DC targeting efficiency (Jahan, Sadat, & Haddadi, 2018). Anti-CD205 monoclonal Ab was specifically chosen to target DCs and led to an in vitro increase in DC maturation as determined by flow cytometry and cytokine release. PLGA polymers were also utilized by Min to capture tumor antigens directly from tumor cells postradiotherapy to upregulate the abscopal effect (Min et al., 2017). The NPs were then able to deliver the captured tumor-specific antigens to APCs and improve treatment efficacy in combination with αPD-1 treatment.

4.2 Liposomes

Liposomes are highly biocompatible, can be readily functionalized, and are able to hold on to their cargo until their lipid bilayer is disrupted. When this does occur, it often causes faster release kinetics than the sustained release of PLGA (Kapadia, Perry, Tian, Luft, & DeSimone, 2015). They are poorly immunogenic on their own, but due to their ease of functionalization can still be effective immunotherapy delivery vehicles (Fontana, Liu, Hirvonen, & Santos, 2017). Lai et al. coated liposomes with a combination of CpG adjuvant and DC targeting mannose to successfully deliver an encapsulated melanoma-specific TRP2 peptide to DCs (Lai et al., 2018). This enhanced DC activation and subsequently increased T cell activation, CD8+ T cells, and interferon-gamma producing cells. In a B16 melanoma model, mice injected with these liposomes displayed inhibition of tumor growth, and prolonged survival. In another use of liposomes as delivery agents, Li et al. loaded a 6-carbon ceramide prodrug into a liposome (LipC6) to enable delivery of an otherwise unusable sphingolipid metabolite due to ceramides aqueous precipitation and cellular impermeability (Li et al., 2018). Ceramide functions as both a chemo- and immunotherapeutic, inducing apoptosis of cancer cells and affecting T-cell signaling. In their hepatocellular carcinoma model, LipC6 injections slowed tumor growth, increased tumor cell apoptosis, and reduced tumor-associated macrophages which suppress the antitumor immune response.

4.3 Dendrimers

Dendrimers offer one of the most highly specific nanoparticle physical characteristics due to their stepwise branching synthesis. With each synthetic generation of synthesis having exactly double the number of surface molecules, particle size and binding sites can be exactly controlled. Polyamidoamine (PAMAM) is one of the most common dendrimers due to its biocompatibility, surface functionalizability, and ease of synthesis. This dendrimer was conjugated with Lewis^b glycopeptides by García-Vallejo to target dendritic cell-specific ICAM-3-grabbing nonintegrin (DC-SIGN) and demonstrated that increasing multimeric presentation increased binding, but beyond generation 5 there was no additional increase in internalization and signaling (García-Vallejo et al., 2013). This resulted in optimal antigen presentation occurring with dendrimers carrying 16–32 glycan units and caused a significant induction of CD4+ and CD8+ T cells. A different polyglutamic dendrimer was modeled computationally by Moura to determine optimal surface functionalization (Moura et al., 2017). This in silico model determined the optimal functionalization of the dendrimer with mannosamine as a targeting adjuvant, and three separate TAAs: MART-1, gp100:44, and gp100:209. This fourth generation dendrimer has 64 carboxylic groups, which they found 32 to be available for covalent bonding to mannosamine but calculated that only 16 mannosamines should be conjugated to leave the other binding groups available for TAAs.

4.4 Virus-Like Particles

VLPs are NPs (20–100 nm) derived from varying viruses which have been inactivated such that they are no longer able to replicate (Buonaguro, Tagliamonte, Tornesello, & Buonaguro, 2011). They are able to present linear and conformational antigens to APCs, resulting in B-cell activation and cross-presentation with both MHC I and II, thereby activating CD4+ and CD8+ T cells (Buonaguro et al., 2011). VLPs can be engineered to target immune cells, augment vaccine efficacy, or loaded with immunogenic ligands via site directed mutagenesis or bioconjugation (Smith, Simon, & Baker, 2013).

While most current VLPs at the clinical stage are directed at viral pathogens, there are a number of emerging indications using VLPs for cancer immunotherapy (Zhang et al., 2015). Bolli et al. developed a VLP (AX09-0M6) targeting the cysteine–glutamate antiporter protein xCT which provides oxidative and chemical insult protection in breast cancer stem cells (Bolli et al., 2018). The VLP produced an oligoclonal antibody response which targeted xCT and increased levels of NK cells in metastatic lung tissue in treated mice. These enabled treated mice to inhibit pulmonary module formulation when challenged with IV $Her2^+$ TUBO-derived tumorsphere injections, or hamper tumor growth in previously established subcutaneous 4T1 tumors. Xi et al. used a combination of Trop-2 as an antigen as it is a commonly overexpressed glycoprotein in cancers, and CD40L as an adjuvant incorporated into the membrane envelope of an HIV gag-based VLP (Xi et al., 2018). When the VLPs were compared with and without CD40L, those incorporating the CD40L demonstrated higher immune responses, reduced tumor growth, and increased survival rate in Lewis tumor bearing mice.

4.5 Inorganic Nanoparticles

Many commonly used pristine metals and other inorganic compounds such as TiO_2, Au, or SiO_2 can also be used to form nanoparticles (Chakraborty, Boer, Selomulya, & Plebanski, 2018). Many of these particles are highly biocompatible, chemically inert, and have tunable surface chemistry allowing them to act as modifiable nanocarriers. There are also issues, however, such as the formation of reactive oxygen species, acting as a catalytic agent of undesirable reactions, inability of metal NPs to biodegrade, or direct cellular toxicity leading to apoptosis of not only tumor cells, but also normal tissue. Fogli et al. functionalized Au and SiO_2 nanoparticles with hepatic and ovarian cancer cell lysates and incubated them with immature DCs (Fogli et al., 2017). The NPs did not cause any toxicity and were taken up by the DCs, but only the Au functionalized NPs promoted DC maturation as observed through upregulation of CD80 and CD83. While both the Au and SiO_2 NPs increased T cell proliferation, SiO_2 NPs only induced CD4+ T cells, while Au NPs induced CD4+ and CD8+ T cell proliferation.

4.6 Hybrid Particles

While nanoparticles offer the ability to encapsulate molecules inside or conjugate them to the surface, there are also some particles which are being joined to each other to create hybrid nanoparticles. Lizotte et al. have previously demonstrated that VLPs derived from CPMV could reduce lung metastasis from existing B16F10 tumors when inhaled (Lizotte et al., 2016). More recently Czapar et al. have used CPMV along with a 4th generation PAMAM dendrimer to form an aggregate hybrid nanoparticle which allowed a sustained release of CPMV (Czapar, Tiu, Veliz, Pokorski, & Steinmetz, 2018). This was used in an ovarian cancer model where CPMV alone required weekly administration, but a single dose of the CPMV-4G dendrimer resulted in similar tumor burden and delayed cancer cell growth.

5. DISCUSSION

Modulating the immune system through immunotherapy methods have advanced cancer therapy. Results from blocking tumor's immunosuppression mechanisms, cancer vaccines, and targeting tumor cells and the tumor microenvironment have been encouraging. It continues to be a promising therapeutic option for cancer but has substantial room for improvement. While nanoparticles have been able to offer significant improvements already, there are numerous opportunities for future advances. With many different types of nanoparticles available as delivery and targeting vehicles, choosing the best type for a specific indication and optimizing that particles properties such as size, shape, charge, material, surface functionalization, and choice of antigen and adjuvant can require a significant amount of work but is critically important. Through these modifications—precise delivery with improved stability, biodistribution, pharmacokinetics, and toxicity profiles—can be achieved as needed for clinical development.

REFERENCES

Bauleth-Ramos, T., Shahbazi, M.-A., Liu, D., Fontana, F., Correia, A., Figueiredo, P., et al. (2017). Nutlin-3a and cytokine co-loaded spermine-modified acetalated dextran nanoparticles for cancer chemo-immunotherapy. *Advanced Functional Materials*, *27*(42), 1703303. https://doi.org/10.1002/adfm.201703303.

Bolli, E., O'Rourke, J. P., Conti, L., Lanzardo, S., Rolih, V., Christen, J. M., et al. (2018). A virus-like-particle immunotherapy targeting epitope-specific anti-xCT expressed on cancer stem cell inhibits the progression of metastatic cancer *in vivo*. *Oncoimmunology*, *7*(3), e1408746. https://doi.org/10.1080/2162402X.2017.1408746.

Bocanegra Gondan, A. I., Ruiz-de-Angulo, A., Zabaleta, A., Gómez Blanco, N., Cobaleda-Siles, B. M., García-Granda, M. J., et al. (2018). Effective cancer immunotherapy in mice by polyIC-imiquimod complexes and engineered magnetic nanoparticles. *Biomaterials*, *170*, 95–115. https://doi.org/10.1016/j.biomaterials.2018.04.003.

Boutros, C., Tarhini, A., Routier, E., Lambotte, O., Ladurie, F. L., Carbonnel, F., et al. (2016). Safety profiles of anti-CTLA-4 and anti-PD-1 antibodies alone and in combination. *Nature Reviews. Clinical Oncology*, *13*(8), 473–486. https://doi.org/10.1038/nrclinonc.2016.58.

Buonaguro, L., Tagliamonte, M., Tornesello, M. L., & Buonaguro, F. M. (2011). Developments in virus-like particle-based vaccines for infectious diseases and cancer. *Expert Review of Vaccines*, *10*(11), 1569–1583. https://doi.org/10.1586/erv.11.135.

Chakraborty, A., Boer, J. C., Selomulya, C., & Plebanski, M. (2018). Amino acid functionalized inorganic nanoparticles as cutting-edge therapeutic and diagnostic agents. *Bioconjugate Chemistry*, *29*(3), 657–671. https://doi.org/10.1021/acs.bioconjchem.7b00455.

Chen, D. S., & Mellman, I. (2013). Oncology meets immunology: The cancer-immunity cycle. *Immunity*, *39*, 1–10. https://doi.org/10.1016/j.immuni.2013.07.012.

Chiavenna, S. M., Jaworski, J. P., & Vendrell, A. (2017). State of the art in anti-cancer mAbs. *Journal of Biomedical Science*, *24*, 15. https://doi.org/10.1186/s12929-016-0311-y.

Christian, D. A., & Hunter, C. A. (2012). Particle-mediated delivery of cytokines for immunotherapy. *Immunotherapy*, *4*(4), 425–441. https://doi.org/10.2217/imt.12.26.

Coley, W. B. (1898). The treatment of inoperable sarcoma with the mixed toxins of erysipelas and Bacillus prodigiosus. *Journal of the American Medical Association*, *XXXI*(8), 389. https://doi.org/10.1001/jama.1898.92450080015001d.

Colzani, B., Pandolfi, L., Hoti, A., Iovene, P. A., Natalello, A., Avvakumova, S., et al. (2018). Investigation of antitumor activities of trastuzumab delivered by PLGA nanoparticles. *International Journal of Nanomedicine*, *13*, 957–973. https://doi.org/10.2147/IJN.S152742.

Cruz, L. J., Rosalia, R. A., Kleinovink, J. W., Rueda, F., Löwik, C. W. G. M., & Ossendorp, F. (2014). Targeting nanoparticles to CD40, DEC-205 or CD11c molecules on dendritic cells for efficient CD8+ T cell response: A comparative study. *Journal of Controlled Release*, *192*, 209–218. https://doi.org/10.1016/J.JCONREL.2014.07.040.

Czapar, A. E., Tiu, B. D. B., Veliz, F. A., Pokorski, J. K., & Steinmetz, N. F. (2018). Slow-release formulation of cowpea mosaic virus for in situ vaccine delivery to treat ovarian cancer. *Advanced Science*, *5*, 1700991. https://doi.org/10.1002/advs.201700991.

De Miguel, D., Gallego-Lleyda, A., Ayuso, J. M., Erviti-Ardanaz, S., Pazo-Cid, R., del Agua, C., et al. (2016). TRAIL-coated lipid-nanoparticles overcome resistance to soluble recombinant TRAIL in non-small cell lung cancer cells. *Nanotechnology*, *27*(18), 185101. https://doi.org/10.1088/0957-4484/27/18/185101.

Demento, S. L., Cui, W., Criscione, J. M., Stern, E., Tulipan, J., Kaech, S. M., et al. (2012). Role of sustained antigen release from nanoparticle vaccines in shaping the T cell memory phenotype. *Biomaterials*, *33*(19), 4957–4964. https://doi.org/10.1016/J.BIOMATERIALS.2012.03.041.

Duan, F., Feng, X., Yang, X., Sun, W., Jin, Y., Liu, H., et al. (2017). A simple and powerful co-delivery system based on pH-responsive metal-organic frameworks for enhanced cancer immunotherapy. *Biomaterials*, *122*, 23–33. https://doi.org/10.1016/J.BIOMATERIALS.2017.01.017.

Dunn, G. P., Bruce, A. T., Ikeda, H., Old, L. J., & Schreiber, R. D. (2002). Cancer immunoediting: From immunosurveillance to tumor escape. *Nature Immunology*, *3*(11), 991–998. https://doi.org/10.1038/ni1102-991.

Dwivedi, P. D., Tripathi, A., Ansari, K. M., Shanker, R., & Das, M. (2011). Impact of nanoparticles on the immune system. *Journal of Biomedical Nanotechnology*, 7(1), 193–194. Retrieved from http://www.ncbi.nlm.nih.gov/pubmed/21485866.

Fogli, S., Montis, C., Paccosi, S., Silvano, A., Michelucci, E., Berti, D., et al. (2017). Inorganic nanoparticles as potential regulators of immune response in dendritic cells. *Nanomedicine*, *12*(14), 1647–1660. https://doi.org/10.2217/nnm-2017-0061.

Fontana, F., Liu, D., Hirvonen, J., & Santos, H. A. (2017). Delivery of therapeutics with nanoparticles: What's new in cancer immunotherapy? *Wiley Interdisciplinary Reviews. Nanomedicine and Nanobiotechnology*, *9*, e1421. https://doi.org/10.1002/wnan.1421.

García-Vallejo, J. J., Ambrosini, M., Overbeek, A., van Riel, W. E., Bloem, K., Unger, W. W. J., et al. (2013). Multivalent glycopeptide dendrimers for the targeted delivery of antigens to dendritic cells. *Molecular Immunology*, *53*(4), 387–397. https://doi.org/10.1016/J.MOLIMM.2012.09.012.

Grimaldi, A. M., Incoronato, M., Salvatore, M., & Soricelli, A. (2017). Nanoparticle-based strategies for cancer immunotherapy and immunodiagnostics. *Nanomedicine*, *12*(19), 2349–2365. https://doi.org/10.2217/nnm-2017-0208.

Guimarães, P. P. G., Gaglione, S., Sewastianik, T., Carrasco, R. D., Langer, R., & Mitchell, M. J. (2018). Nanoparticles for immune cytokine TRAIL-based cancer therapy. *ACS Nano*, *12*(2), 912–931. https://doi.org/10.1021/acsnano.7b05876.

Hansel, T. T., Kropshofer, H., Singer, T., Mitchell, J. A., & George, A. J. T. (2010). The safety and side effects of monoclonal antibodies. *Nature Reviews. Drug Discovery*, *9*(4), 325–338. https://doi.org/10.1038/nrd3003.

Hassan, H. A. F. M., Smyth, L., Wang, J. T.-W., Costa, P. M., Ratnasothy, K., Diebold, S. S., et al. (2016). Dual stimulation of antigen presenting cells using carbon nanotube-based vaccine delivery system for cancer immunotherapy. *Biomaterials*, *104*, 310–322. https://doi.org/10.1016/j.biomaterials.2016.07.005.

Hercus, T. R., Thomas, D., Guthridge, M. A., Ekert, P. G., King-Scott, J., Parker, M. W., et al. (2009). The granulocyte-macrophage colony-stimulating factor receptor: Linking its structure to cell signaling and its role in disease. *Blood*, *114*(7), 1289–1298. https://doi.org/10.1182/blood-2008-12-164004.

Hobo, W., Novobrantseva, T. I., Fredrix, H., Wong, J., Milstein, S., Epstein-Barash, H., et al. (2013). Improving dendritic cell vaccine immunogenicity by silencing PD-1 ligands using siRNA-lipid nanoparticles combined with antigen mRNA electroporation. *Cancer Immunology, Immunotherapy*, *62*(2), 285–297. https://doi.org/10.1007/s00262-012-1334-1.

Irvine, D. J., Hanson, M. C., Rakhra, K., & Tokatlian, T. (2015). Synthetic nanoparticles for vaccines and immunotherapy. *Chemical Reviews*, *115*(19), 11109–11146. https://doi.org/10.1021/acs.chemrev.5b00109.

Jahan, S. T., Sadat, S. M., & Haddadi, A. (2018). Design and immunological evaluation of anti-CD205-tailored PLGA-based nanoparticulate cancer vaccine. *International Journal of Nanomedicine*, *13*, 367–386. https://doi.org/10.2147/IJN.S144266.
Kang, T., Huang, Y., Zhu, Q., Cheng, H., Pei, Y., Feng, J., et al. (2018). Necroptotic cancer cells-mimicry nanovaccine boosts anti-tumor immunity with tailored immune-stimulatory modality. *Biomaterials*, *164*, 80–97. https://doi.org/10.1016/j.biomaterials.2018.02.033.
Kapadia, C. H., Perry, J. L., Tian, S., Luft, J. C., & DeSimone, J. M. (2015). Nanoparticulate immunotherapy for cancer. *Journal of Controlled Release*, *219*, 167–180. https://doi.org/10.1016/J.JCONREL.2015.09.062.
Khalil, D. N., Smith, E. L., Brentjens, R. J., & Wolchok, J. D. (2016). The future of cancer treatment: Immunomodulation, CARs and combination immunotherapy. *Nature Reviews. Clinical Oncology*, *13*(5), 273–290. https://doi.org/10.1038/nrclinonc.2016.25.
Kim, H., Niu, L., Larson, P., Kucaba, T. A., Murphy, K. A., James, B. R., et al. (2018). Polymeric nanoparticles encapsulating novel TLR7/8 agonists as immunostimulatory adjuvants for enhanced cancer immunotherapy. *Biomaterials*, *164*, 38–53. https://doi.org/10.1016/J.BIOMATERIALS.2018.02.034.
Kranz, L. M., Diken, M., Haas, H., Kreiter, S., Loquai, C., Reuter, K. C., et al. (2016). Systemic RNA delivery to dendritic cells exploits antiviral defence for cancer immunotherapy. *Nature*, *534*(7607), 396–401. https://doi.org/10.1038/nature18300.
Kuai, R., Ochyl, L. J., Bahjat, K. S., Schwendeman, A., & Moon, J. J. (2017). Designer vaccine nanodiscs for personalized cancer immunotherapy. *Nature Materials*, *16*(4), 489–496. https://doi.org/10.1038/nmat4822.
Lai, C., Duan, S., Ye, F., Hou, X., Li, X., Zhao, J., et al. (2018). The enhanced antitumor-specific immune response with mannose- and CpG-ODN-coated liposomes delivering TRP2 peptide. *Theranostics*, *8*(6), 1723–1739. https://doi.org/10.7150/thno.22056.
Li, G., Liu, D., Kimchi, E. T., Kaifi, J. T., Qi, X., Manjunath, Y., et al. (2018). Nanoliposome C6-ceramide increases the anti-tumor immune response and slows growth of liver tumors in mice. *Gastroenterology*, *154*(4) 1024–1036.e9. https://doi.org/10.1053/J.GASTRO.2017.10.050.
Liang, R., Xie, J., Li, J., Wang, K., Liu, L., Gao, Y., et al. (2017). Liposomes-coated gold nanocages with antigens and adjuvants targeted delivery to dendritic cells for enhancing antitumor immune response. *Biomaterials*, *149*, 41–50. https://doi.org/10.1016/j.biomaterials.2017.09.029.
Liu, L., Yi, H., Wang, C., He, H., Li, P., Pan, H., et al. (2016). Integrated nanovaccine with microRNA-148a inhibition reprograms tumor-associated dendritic cells by modulating miR-148a/DNMT1/SOCS1 axis. *Journal of Immunology (Baltimore, Md.: 1950)*, *197*(4), 1231–1241. https://doi.org/10.4049/jimmunol.1600182.
Lizotte, P. H., Wen, A. M., Sheen, M. R., Fields, J., Rojanasopondist, P., Steinmetz, N. F., et al. (2016). In situ vaccination with cowpea mosaic virus nanoparticles suppresses metastatic cancer. *Nature Nanotechnology*, *11*(3), 295–303. https://doi.org/10.1038/nnano.2015.292.
Mi, Y., Smith, C. C., Yang, F., Qi, Y., Roche, K. C., Serody, J. S., et al. (2018). A dual immunotherapy nanoparticle improves T-cell activation and cancer immunotherapy. *Advanced Materials*, e1706098. https://doi.org/10.1002/adma.201706098.
Min, Y., Roche, K. C., Tian, S., Eblan, M. J., McKinnon, K. P., Caster, J. M., et al. (2017). Antigen-capturing nanoparticles improve the abscopal effect and cancer immunotherapy. *Nature Nanotechnology*, *12*(9), 877–882. https://doi.org/10.1038/nnano.2017.113.
Moura, L. I. F., Martinho, N., Silva, L. C., Barata, T. S., Brocchini, S., Florindo, H. F., et al. (2017). Poly-glutamic dendrimer-based conjugates for cancer vaccination—A computational design for targeted delivery of antigens. *Journal of Drug Targeting*, *25*(9–10), 873–880. https://doi.org/10.1080/1061186X.2017.1363213.

Oberli, M. A., Reichmuth, A. M., Dorkin, J. R., Mitchell, M. J., Fenton, O. S., Jaklenec, A., et al. (2017). Lipid nanoparticle assisted mRNA delivery for potent cancer immunotherapy. *Nano Letters*, *17*(3), 1326–1335. https://doi.org/10.1021/acs.nanolett.6b03329.

Pardoll, D. M. (2012). The blockade of immune checkpoints in cancer immunotherapy. *Nature Reviews. Cancer*, *12*(4), 252–264. https://doi.org/10.1038/nrc3239.

Patel, S. A., & Minn, A. J. (2018). Combination cancer therapy with immune checkpoint blockade: Mechanisms and strategies. *Immunity*, *48*(3), 417–433. https://doi.org/10.1016/j.immuni.2018.03.007.

Qiu, H., Min, Y., Rodgers, Z., Zhang, L., & Wang, A. Z. (2017). Nanomedicine approaches to improve cancer immunotherapy. *Wiley Interdisciplinary Reviews: Nanomedicine and Nanobiotechnology*, *9*(5), 1–15. https://doi.org/10.1002/wnan.1456.

Ray, M., Lee, Y.-W., Hardie, J., Mout, R., Yeşilbag Tonga, G., Farkas, M. E., et al. (2018). CRISPRed macrophages for cell-based Cancer immunotherapy. *Bioconjugate Chemistry*, *29*(2), 445–450. https://doi.org/10.1021/acs.bioconjchem.7b00768.

Rosenberg, S. A. (2014). IL-2: The first effective immunotherapy for human cancer. *Journal of Immunology (Baltimore, Md.: 1950)*, *192*(12), 5451–5458. https://doi.org/10.4049/jimmunol.1490019.

Sahin, U., Karikó, K., & Türeci, Ö. (2014). mRNA-based therapeutics—Developing a new class of drugs. *Nature Reviews. Drug Discovery*, *13*(10), 759–780. https://doi.org/10.1038/nrd4278.

Schreiber, R. D., Old, L. J., & Smyth, M. J. (2011). Cancer immunoediting: Integrating immunity's roles in cancer suppression and promotion. *Science*, *331*(6024), 1565–1570. https://doi.org/10.1126/science.1203486.

Silva, J. M., Videira, M., Gaspar, R., Préat, V., & Florindo, H. F. (2013). Immune system targeting by biodegradable nanoparticles for cancer vaccines. *Journal of Controlled Release*, *168*(2), 179–199. https://doi.org/10.1016/J.JCONREL.2013.03.010.

Smith, D. M., Simon, J. K., & Baker, J. R. (2013). Applications of nanotechnology for immunology. *Nature Reviews. Immunology*, *13*(8), 592–605. https://doi.org/10.1038/nri3488.

Tang, J., Chen, Z., Sun, B., Dong, J., Liu, J., Zhou, H., et al. (2016). Polyhydroxylated fullerenols regulate macrophage for cancer adoptive immunotherapy and greatly inhibit the tumor metastasis. *Nanomedicine: Nanotechnology, Biology and Medicine*, *12*(4), 945–954. https://doi.org/10.1016/J.NANO.2015.11.021.

Topalian, S. L., Drake, C. G., & Pardoll, D. M. (2012). The blockade of immune checkpoints in cancer immunotherapy. *Cancer Cell*, *27*(4), 252–264. https://doi.org/10.1038/nrc3239.

van der Burg, S. H., Arens, R., Ossendorp, F., van Hall, T., & Melief, C. J. M. (2016). Vaccines for established cancer: Overcoming the challenges posed by immune evasion. *Nature Reviews. Cancer*, *16*(4), 219–233. https://doi.org/10.1038/nrc.2016.16.

Vasievich, E. A., Ramishetti, S., Zhang, Y., & Huang, L. (2012). Trp2 peptide vaccine adjuvanted with (*R*)-DOTAP inhibits tumor growth in an advanced melanoma model. *Molecular Pharmaceutics*, *9*(2), 261–268. https://doi.org/10.1021/mp200350n.

Xi, W., Ke, D., Min, L., Lin, W., Jiahui, Z., Fang, L., et al. (2018). Incorporation of CD40 ligand enhances the immunogenicity of tumor-associated calcium signal transducer 2 virus-like particles against lung cancer. *International Journal of Molecular Medicine*, *41*, 3671–3679. https://doi.org/10.3892/ijmm.2018.3570.

Xie, Y.-Q., Wei, L., & Tang, L. (2018). Immunoengineering with biomaterials for enhanced cancer immunotherapy. *Wiley Interdisciplinary Reviews: Nanomedicine and Nanobiotechnology*, e1506. https://doi.org/10.1002/wnan.1506 (September 2017).

Yan, W.-L., Shen, K.-Y., Tien, C.-Y., Chen, Y.-A., & Liu, S.-J. (2017). Recent progress in GM-CSF-based cancer immunotherapy. *Immunotherapy*, *9*(4), 347–360. https://doi.org/10.2217/imt-2016-0141.

Yin, H., Kanasty, R. L., Eltoukhy, A. A., Vegas, A. J., Dorkin, J. R., & Anderson, D. G. (2014). Non-viral vectors for gene-based therapy. *Nature Reviews. Genetics*, *15*(8), 541–555. https://doi.org/10.1038/nrg3763.

Zhang, P., Chen, Y., Zeng, Y., Shen, C., Li, R., Guo, Z., et al. (2015). Virus-mimetic nanovesicles as a versatile antigen-delivery system. *Proceedings of the National Academy of Sciences of the United States of America*, *112*(45), E6129–E6138. https://doi.org/10.1073/pnas.1505799112.

Zhuang, X., Wu, T., Zhao, Y., Hu, X., Bao, Y., Guo, Y., et al. (2016). Lipid-enveloped zinc phosphate hybrid nanoparticles for codelivery of H-2K b and H-2D b -restricted antigenic peptides and monophosphoryl lipid A to induce antitumor immunity against melanoma. *Journal of Controlled Release*, *228*, 26–37. https://doi.org/10.1016/j.jconrel.2016.02.035.

Zolnik, B. S., González-Fernández, A., Sadrieh, N., & Dobrovolskaia, M. A. (2010). Nanoparticles and the immune system. *Endocrinology*, *151*(2), 458–465. https://doi.org/10.1210/en.2009-1082.

CHAPTER THREE

Ultrasound Contrast Agents and Delivery Systems in Cancer Detection and Therapy

Al de Leon*, Reshani Perera*, Pinunta Nittayacharn[†], Michaela Cooley*, Olive Jung*[,‡], Agata A. Exner*[,†,1]

*Department of Radiology, Case Western Reserve University, Cleveland, OH, United States
[†]Department of Biomedical Engineering, Case Western Reserve University, Cleveland, OH, United States
[‡]Nuffield Department of Medicine, University of Oxford, Oxford, United Kingdom
[1]Corresponding author: e-mail address: agata.exner@case.edu

Contents

Abstract

Ultrasound is the second most utilized imaging modality in the world because it is widely accessible, robust, and safe. Aside from its extensive use in diagnostic imaging, ultrasound has also been frequently utilized in therapeutic applications. Particularly, when combined with appropriate delivery systems, ultrasound provides a flexible platform for simultaneous real-time imaging and triggered release, enabling precise, on-demand drug delivery to target sites. This chapter will discuss the basics of ultrasound including its mechanism of action and how it can be used to trigger the release of encapsulated drug either through thermal or cavitation effects. Fundamentals of ultrasound contrast agents, how they enhance ultrasound signals, and how they can be modified to function as carriers for triggered and targeted release of drugs will also be discussed.

Advances in Cancer Research, Volume 139
ISSN 0065-230X
https://doi.org/10.1016/bs.acr.2018.04.002

1. INTRODUCTION

Medical ultrasound imaging, also referred to as diagnostic sonography or ultrasonography, involves dynamic imaging of internal body structures based on how their interfaces reflect the projected beam of high-frequency sound. Ultrasound is considered sound with a frequency greater than 20,000 Hz, which is above the range of human hearing, with most diagnostic applications using frequencies in the 3–20 MHz range (Chemat et al., 2017; Sancheti & Gogate, 2017; Yu & Hossack, 2017). Ultrasound imaging utilizes a piezoelectric transducer, which projects a beam of ultrasound that propagates through the body until it reaches a tissue, where it is reflected, absorbed, or transmitted. The transmitted sound wave will continue to propagate until it reaches another interface; this cycle continues until all projected sound wave is either completely absorbed or reflected. The reflected sound wave is then collected by the same transducer and an image is produced by analyzing the strength (intensity) of the reflected wave and the time it takes for it to travel through the body.

Historically, ultrasound concepts were mainly utilized for purposes outside of medicine. In the 1930–1940s, Karl Dussik was among the first physicians who used ultrasound for medical diagnostic purposes when he, along with his physicist brother, attempted to perform an intracranial tumor diagnosis (Stouffer, 2004). Although the first attempt was unsuccessful due to the presence of artifacts, they managed to produce images of intracranial regions and diagnose tumors by 1942. Following the first report, it was not until the late 1960s when real-time ultrasound imaging, similar to modern ultrasound techniques, gained widespread clinical usage. Over the next several decades, ultrasound modes such as pulsed-wave Doppler, color imaging, and harmonic imaging were successfully developed and optimized. Today, clinical ultrasound is an easily transportable device with increasing applicability and constant evolution into what will be the future of medical imaging.

Between 2000 and 2011, ultrasound was the second most used medical imaging modality in the world after X-ray as shown in Fig. 1 (Szabo, 2013). Currently, common diagnostic applications of ultrasound imaging include (i) gynecological applications for identifying fetal abnormalities; (ii) visualizing soft tissues such as the heart, cysts, muscle, tendons, and blood vessels; and (iii) specialized probing for gastrointestinal organs such as the esophagus and intestines (Fenster & Downey, 1996; Lindner, 2004; Tsui & Suresh, 2010). Consequently, ultrasound imaging is one of the

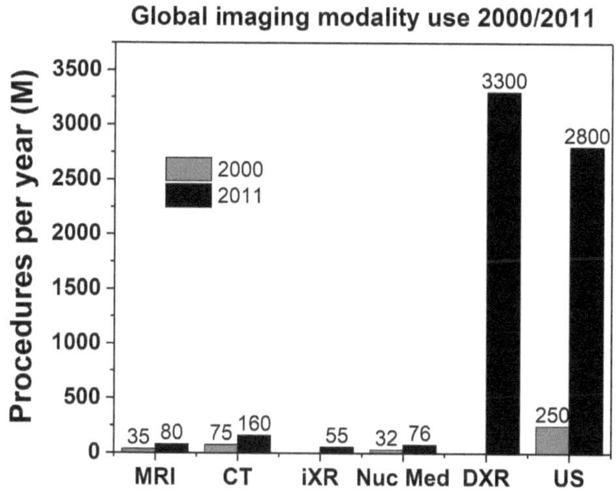

Fig. 1 Estimated number of imaging exams performed worldwide for years 2000 and 2011. *CT*, computed tomography; *DXR*, digital X-ray; *iXR*, interventional X-ray; *MRI*, magnetic resonance imaging; *Nuc Med*, nuclear medicine; *US*, diagnostic ultrasound (Szabo, 2013).

most important imaging modalities in cardiology, urology, and ophthalmology, among many other medical specialties. Ultrasound has also been used for therapeutic applications, particularly for bone healing, sonophoresis, sonoporation, gene therapy, sonothrombolysis, and physiotherapy (Baker, Robertson, & Duck, 2001; Robertson & Baker, 2001). Application of ultrasound to a tissue induces stable cavitation, due to pressure changes, that has been shown to improve healing by increasing blood flow between the injured tissue and surrounding tissues (Young & Dyson, 1990). Furthermore, the positive thermal effects caused by oscillations due to ultrasound can relieve muscle spasms. High-intensity focused ultrasound (HIFU) is also used for purposeful tissue destruction. It has been applied to cancer therapy for destruction of solid tumors, treatment of tremors in Parkinson's disease, and cosmetically, as an alternative to liposuction (Magara et al., 2014). Numerous studies have been conducted on the effect of HIFU-based imaging and treatment of prostate cancer, and the results have been promising. Its use has also been explored in theranostics of pancreatic and liver cancer (Yasuda, Mukai, Fujimoto, Nakajima, & Kawai, 1988).

The following sections of this chapter will discuss how ultrasound has been used in the field of theranostics to visualize the journey of various drug-loaded constructs, quantify biodistribution and kinetics of these agents, and to trigger the release of therapeutic agents at the target site. The enhancement of ultrasound images by using ultrasound contrast agents (UCAs), effect of its chemical components, mechanical properties, and their behavior under ultrasound field will also be covered. The use of ultrasound to trigger release of drug via thermal and cavitation effects, along with various drug delivery systems designed to be triggered by ultrasound, will be discussed in detail. Lastly, the use of UCA for simultaneous imaging and delivery of drugs will be examined.

2. BENEFITS OF ULTRASOUND

Among the commonly utilized imaging tools available to physicians and researchers, such as computed tomography (CT), positron emission tomography (PET), magnetic resonance imaging (MRI), single-photon emission computed tomography (SPECT), and optical imaging, ultrasound continues to be the most convenient, cost effective, and safe (Frangioni, 2008; Sistrom & McKay, 2005). Another significant advantage of medical ultrasound is that it can provide real-time (15–20 frames per second, or higher) images without exposing patients to repeated doses of radiation (when, for example, rapid repeated image acquisition is performed with CT). This enables unique ultrasound applications ranging from image-guided biopsy and injections to rapid analysis of the kinetics of echogenic agents. Consequently, ultrasound is often one of the first options taken by researchers and physicians to evaluate, diagnose, and treat patients (Calliada, Campani, Bottinelli, Bozzini, & Sommaruga, 1998; Stouffer, 2004).

3. ULTRASOUND CONTRAST AGENTS

The development of UCAs facilitated a period of significant growth for ultrasound imaging. UCAs are biocolloids that enhance the intensity of the backscattered echo. The enhancement is proportional to the change in acoustic impedance between internal body structures and UCAs (Goldberg, Liu, & Forsberg, 1994). An UCA acts as an echo-enhancer because it undergoes a much higher degree of compression and expansion under an ultrasound field when compared to the surrounding blood or tissues. In addition, UCA under an ultrasound field resonates at a frequency that is a combination of its resonant frequency (fundamental) and its harmonics (multiples of the fundamental) (Calliada et al., 1998; Dejong, Tencate, Lancee, Roelandt, & Bom, 1991). This is in contrast with the internal body structures that resonate mostly at their fundamental frequency under an

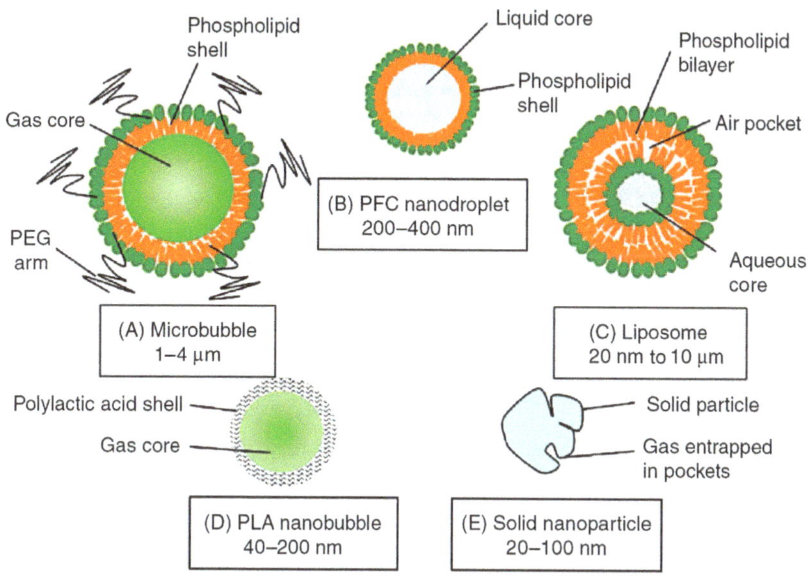

Fig. 2 Different types of ultrasound contrast agents (Oláh, 2016).

ultrasound field. Several types of UCAs have been developed (Fig. 2) for both clinical and research applications, including microbubbles and nanobubbles (gas–liquid emulsions), nanodrops/nanoemulsions (liquid–liquid emulsions), and echogenic liposomes (liquid–liquid emulsion with a bilayer membrane). Unless otherwise indicated, the following discussions will be focused on both microbubbles and nanobubbles as UCAs.

Clinical utility of UCAs was first discovered in the late 1960s when saline bubbles were accidentally introduced during echocardiography of the aortic root (Calliada et al., 1998). This first UCA, however, was not stable and was too transient to be utilized extensively in a clinical setting. Consequently, significant effort was devoted to developing highly stable UCA. The difficulty in creating such stable UCA was mostly attributed to the efficient filtering of lung capillaries. Numerous formulations were tested, but it was not until 1990s that the left heart chambers could be contrasted with UCA (Ignee, Atkinson, Schuessler, & Dietrich, 2016). The first FDA-approved UCA, Echovist® (approved in 1991), was a simple echo-enhancer consisting of micron-sized air bubbles stabilized by galactose that was primarily used for shunt imaging in cardiology and hysterosalpingo-contrast sonography. The first-generation blood pool agent Albunex® (comprised of air-filled microbubbles stabilized by albumin) was approved in 1995. Both agents, while useful during the aforementioned clinical procedures, did not provide sufficient time for proper real-time imaging due to their relative instability compared to the more recent UCA (Albunex was found to only last <30 s in the left ventricle) (Feinstein et al., 1990; Ignee et al., 2016). In formulating UCA for real-time monitoring, much of the subsequent efforts were shifted toward replacing the core from air to less blood-soluble gases such as perfluorocarbons (which prolong circulation time) and improving the acoustic parameters that can increase signal-to-noise ratio and signal sensitivity (Paefgen, Doleschel, & Kiessling, 2015).

Notably, development and application of UCA in the United States were heavily stalled by a "black box warning" issued by US Food and Drug Administration (FDA) in 2007 after patients have reported serious adverse effects following UCA administration. However, the contraindications were downgraded to warnings, and subsequent safety studies in >200,000 patients receiving Optison® and Definity® showed very low adverse effects. There was only one case of acute anaphylactoid reaction out of 10,000 patients who have been given the said UCA (Aggeli et al., 2008; Dolan et al., 2009; Main et al., 2008). In response to the persistently proven safety and the applicability of the agents, there has been a continued rise in interest and research toward these ultrasound imaging agents. The most recently approved FDA UCA includes Lumason®, which is now used for liver lesion imaging. Below is a list of UCAs that are approved by the FDA for commercial purposes (Dawson, 2002; Perera et al., 2015).

3.1 Interactions of UCA With Ultrasound

As previously mentioned, UCAs act as echo-enhancers, in which the intensity of the backscattered echo depends on the change in acoustic impedance between the blood and the core of the agents, compressibility and density of the agents, and property of its surrounding microenvironment (Klibanov, Shevchenko, Raju, Seip, & Chin, 2010; Qin, Caskey, & Ferrara, 2009). The size and chemical components of UCA dictate how they respond under ultrasound. As an illustration, two microbubbles of the same diameter can have a 1000-fold difference in echogenicity depending on the gas and the property of the membrane encapsulating the gas core (Calliada et al., 1998; Klibanov et al., 2010; Qin et al., 2009).

In the presence of ultrasound, UCAs resonate at a frequency comprised of their resonant and harmonic frequencies. The second harmonic frequency is usually the only harmonic besides the fundamental frequency that is strong enough to be collected and analyzed (Dejong et al., 1991). As mentioned earlier, the presence of harmonic oscillation of UCA is a crucial differentiation between the UCA and the surrounding tissue. While the UCAs are able to resonate with harmonic frequencies, the adjacent tissues do not resonate or have very little harmonic oscillations. Accordingly imaging techniques that take advantage of this discrepancy, such as "contrast harmonic imaging," can be applied to significantly increase the signal-to-noise ratio of bubble signal, giving the technique superb sensitivity. Peak rarefactional pressure (PRP), which is the time period of negative half-cycle of pressure wave, also plays an important role in the contrast enhancement of ultrasound (Apfel & Holland, 1991; Mehier-Humbert, Bettinger, Yan, & Guy, 2005). With a higher PRP, the oscillation of the agents—their expansion and collapse—has a nonlinear relationship to the driving pressure (Qin et al., 2009). For higher values of transmitted PRP, the expansion is followed by a comparatively quicker collapse (Qin et al., 2009; Wrenn et al., 2012). These phenomena have also been utilized in image acquisition and processing algorithms.

3.2 Microbubbles as UCAs

There are a number of UCAs available for commercial and research purposes; however, microbubbles (MB) and nanobubbles (NB) are of interest because of their low cost, ease of preparation, and robustness. Both MB and NB are small gas-filled bubbles with diameters larger and smaller than 1 μm, respectively (Agarwal, Ng, & Liu, 2011). The introduction of MB into clinical settings drastically improved the imaging of various organs and management of various diseases. There are a number of commercially available MB approved by FDA for clinical use as shown in Table 1. They have been clinically used to image heart, liver, and veins, among others. Although MB provide excellent contrast, they suffers from some disadvantages such as stability. MB tend to gradually decrease in size and collapse due to dissolution of their inner gas into the surrounding aqueous environment. In addition, imaging with MB is limited to vasculature because of their size. These limitations prompt the need to reduce the size of the UCA down to nanolevel.

3.3 Nanobubbles as UCAs

Although all commercially available UCAs are MB, NB have recently become the focus of most research on contrast imaging, specifically for cancer imaging, because their size facilitates their movement through small

Table 1 Commercially Available Ultrasound Contrast Agents Approved by FDA

UCA	Year Approved	Components	Clinical Use
Echovist®	1991	Galactose-stabilized air bubbles	Shunt imaging (cardiology)
Albunex®	1995	Albumin-stabilized air bubbles	Cardiology
Levovist®	1995	Galactose- and palmitic acid-stabilized air bubbles	Cardiology, liver imaging
Optison®	1997	Albumin-stabilized perflutren bubbles	Left ventricle contrast imaging (opacification)
SonoVue®	2001 (approved by FDA as Lumason® in 2014)	Phospholipid-stabilized sulfur hexafluoride bubbles	Similar to Optison; also for breast, liver, portal vein, extracranial carotid and peripheral arteries
Definity®	2001	Phospholipid-stabilized octafluoropropane bubbles	Left ventricle contrast imaging (opacification)

vascular endothelial gaps in defective tumor blood vessels to reach the tumor parenchyma. Vascular endothelial gaps are typically too small for MB to pass through in order to reach the tumor that has a larger gap (Agarwal et al., 2011). Cai et al., Wu et al., and Gao et al. among others have shown that while the image enhancement of NB in vitro was similar to that of microbubbles such as SonoVue® or Definity®, NB had longer tumor enhancement time compared to MB in in vivo experiments (Cai et al., 2015; Rapoport et al., 2011; Wu et al., 2013). This enhancement was attributed to the incremental accumulation of signal as more NB extravasate to the vasculature of the tumor. They also found that the addition of DSPE-PEG2000 lipids increased its retention by preventing NB from being cleared by the reticuloendothelial system. Another advantage brought about by the small size of NB is that they do not get trapped in the blood pool after intravenous injection. In contrast, MB usually pool in the bloodstream and do not circulate as much as desired, which is their main limitation for tumor-targeted imaging (Agarwal et al., 2011). Also, NB tend to remain in their native form for a longer time. It was found that NB retain hydrogen bonds similar to those in ice and gas hydrates, which allows for reduced diffusivity and balance against high internal Laplace pressure (Agarwal et al., 2011). Furthermore, surface charge, surface tension, and membrane crystallinity all contribute to the observed stability of NB.

3.4 Chemical Components of Microbubbles and Nanobubbles

MB and NB are commonly composed of phospholipids, such as phosphatidylcholine, phosphatidylethanolamine, and phosphatidylglycerol, and gas such as air, sulfur hexafluoride, and fluorocarbon (Fig. 3). Depending on the application, other chemical components may also be incorporated. These include ligands and antibodies for targeting sites of interest in vivo; cross-linkers to improve stability and decrease diffusivity of the gas core; edge activators to manipulate the mechanical properties of the bubble membrane; and, drugs that can be delivered by using intense ultrasound signal (Agarwal et al., 2011; Dawson, 2002; Perera et al., 2014). Polymers such as poly(lactic acid) (PLA), poly(ε-caprolactone) (PCL), poly(glycolic acid), poloxamer (Pluronics), and poly(lactic-co-glycolic acid) (PLGA) have also been used to stabilize the gas core due to their consistency in shell composition and biodegradability (Agarwal et al., 2011). Poly(ethylene glycol) (PEG) is also frequently introduced on the surface to reduce immunogenicity and increase the bubbles' retention in the vasculature.

Variation on chemical components of UCA, acoustic pressure, fluid viscosity, ambient temperature and pressure, gas diffusivity, and size distribution can all affect echogenicity and stability of UCA. Shekhar et al. further proposed that shell properties and stability of UCA are highly dependent on

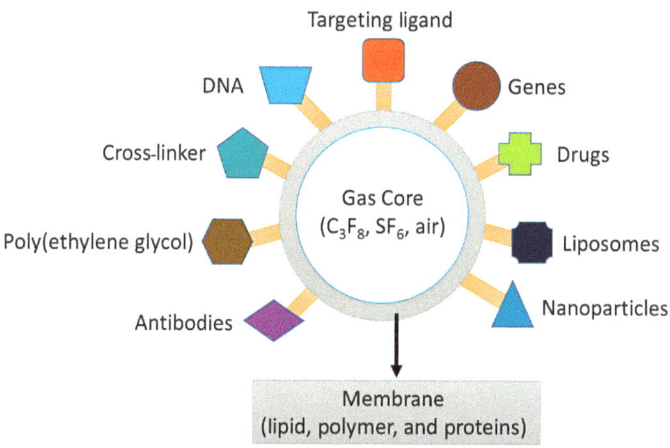

Fig. 3 Schematic of possible components of MB and NB as UCAs.

temperature, and shell properties in turn affect the threshold of nonlinear bubble activity (Shekhar, Smith, Raymond, & Holland, 2018). Bader and Holland further proved that UCA destruction thresholds are directly correlated to their shell properties (Bader & and Holland, 2013). Most recently, it was shown that 1,2-distearoyl-*sn*-glycero-3-phosphocholine (DSPC) and 1,2-dipalmitoyl-rac-glycero-3-phospho (dimethylaminoethanol)-methoxy polyethylene glycol 5000 (DPPE-MPEG5000) could be utilized to resist the size and shape deformation of MB during gas exchange, which allows for longer retention of the encapsulate gas (Bader & and Holland, 2013). The same group showed that compared to pure oxygen in the gas core, combination of oxygen and fluorocarbon gas significantly hindered the dissolution of MB due to the fluorocarbon's low aqueous solubility and diffusivity. Lastly, Perera et al. and Hernandez et al. were able to show that NB formulated with an interpenetrating network of *N*,*N*-diethylacrylamide (NNDEA) are more stable than noncross-linked NB under ultrasound field in vivo and in vitro (Hernandez, Gulati, Fioravanti, Stewart, & Exner, 2017).

4. ULTRASOUND AS AN EXTERNAL, NONINVASIVE TRIGGER FOR DRUG DELIVERY

Ultrasound is a powerful imaging modality that can also be applied to noninvasive on-demand drug delivery (Mitragotri, 2005; Oerlemans et al., 2010; Park, 2010). It has been used for research in the delivery of several therapeutic agents, including chemotherapeutic agents, proteins, and genetic materials encapsulated in stimuli responsive carriers. The encapsulation in a bubble, as in any other delivery system, prevents the drug from interacting with surrounding healthy cells as it circulates throughout the body. In this case, however, once the carrier reaches the desired location, an ultrasound field can be applied as an external stimulus to trigger the release of the drug (Couture, Foley, Kassell, Larrat, & Aubry, 2014). The drug release can be via either two mechanisms: destruction of the carrier to release the drug physically trapped, or breaking the chemical bond between the carrier and the drug (Couture et al., 2014). Ultrasound as a triggering mechanism can be high frequency (>200 kHz) or low frequency (20–200 kHz) depending on the target application, and the application of ultrasound can result in thermal effects, cavitation, or a combination of both (Fig. 4A).

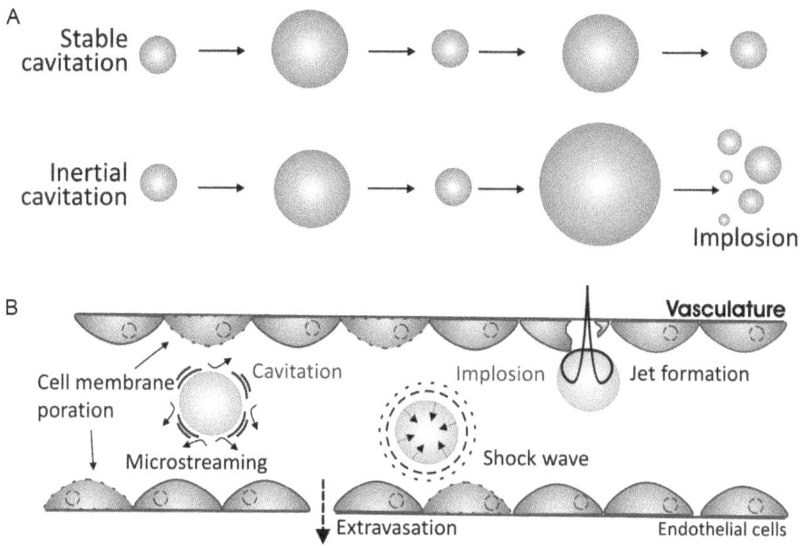

Fig. 4 Schematic representation of (A) stable and inertial cavitation of microbubbles and (B) biological effects which may arise during cavitation and implosion of microbubbles (Lentacker, De Smedt, & Sanders, 2009).

4.1 Thermal Effects of Ultrasound for Triggered Release

High-intensity ultrasonic waves induce oscillation of materials within the ultrasound field. This mechanical oscillation is then converted into thermal energy, which increases the temperature within the treatment zone (Couture et al., 2014). This usually occurs when focused ultrasound (FUS) or HIFU transducers set at moderate acoustic pressure, high duty cycles, and long treatment times are used. In order to avoid unintended damage on surrounding cells due to prolonged heating, drug carriers are designed so that they release their payload at temperature a few degrees above the physiological temperature (e.g., 42–43°C) (Couture et al., 2014). For example, Doxil®, an FDA-approved agent, is a temperature-sensitive liposome that has been shown to release 80% of encapsulated Doxorubicin (Dox) in vitro under 15 min of exposure to ultrasound at 43°C (Phenix, Togtema, Pichardo, Zehbe, & Curiel, 2014).

4.2 Cavitation Effect of Ultrasound for Triggered Release

Cavitation occurs when bubbles rapidly grow and collapse in milliseconds as a result of alternating high and low pressure phases of the ultrasound wave (Barnett et al., 1994). Cavitation effect is considered to be more important than thermal effect for drug delivery and release (Pitt, Husseini, & Staples, 2004). Cavitation can either be inertial or noninertial (stable) as shown in Fig. 4. Noninertial cavitation occurs when bubbles oscillate in a stable manner due to a low-amplitude ultrasound field. The size of the bubble varies just around its resting or equilibrium size, and microstreams of liquid surrounding the bubble start to form. These microstreams cause shearing that releases the encapsulated drugs and forms transient openings through the cell membrane, which subsequently results in an influx of released drug into the cells (Sboros, 2008). Inertial cavitation, on the other hand, occurs when the intensity of the applied ultrasound is high enough that it causes the bubbles to oscillate in an unstable manner, expanding larger than their critical size and collapsing into smaller bubbles. This collapse generates shock waves and microjets (Fig. 4B), which rupture the carrier to release the loaded drug, sonoporation of the cell membrane, and permeabilization of blood vessels (Phenix et al., 2014). Sonoporation of the cell membrane promotes the intracellular delivery of delivered drugs, while permeabilization of blood vessels induces extravasation of drug-loaded carriers. The mechanical stimulation of blood vessel caused by cavitation can also enhance the permeability of the blood–brain barrier (BBB) (Burgess, Shah, Hough, & Hynynen, 2015). The intensity of the ultrasound required to open the BBB is significantly reduced when combined with gas bubbles. This makes the transcranial ultrasound treatment feasible by improving the efficiency of drug or gene delivery for neurological diseases (Hynynen, 2010; Kovacs et al., 2017).

4.3 Drug Delivery Systems Triggered by Ultrasound

Aside from bubbles, other systems, such as polymeric micelles, liposomes, polyelectrolyte microcontainers, or multilayered capsules, can also be used as drug carriers that can be triggered by ultrasound (Lee, Lin, Chen, & Thomas, 2008; Lin & Thomas, 2003; Rapoport, Herron, Pitt, & Pitina, 1999; Shchukin, Gorin, & Möhwald, 2006). Kost et al. was the first to propose the concept of releasing drugs by eroding the polymer carrier by ultrasound-induced cavitation (Kost, Leong, & Langer, 1989). When micelles are used as the drug carrier, there are two events that can result when ultrasound is used for triggered release: (i) reversible release of drug as a result of physical disruption of micelles, or (ii) irreversible release of drug as a result of chemical disruption of micelles. Although ultrasound can cause physical disruption of micelles while it is being applied, once the ultrasound is stopped, the micelles can reassemble and the drug can be reencapsulated; thus, the drug release is reversible (Husseini & Pitt, 2008; Sirsi & Borden, 2014).

The effect of ultrasound intensity, frequency, and pulse on the release of drug from micelles has been comprehensively studied. Husseini et al. encapsulated Dox in Pluronic® micelles and its release was triggered by ultrasound (Husseini, Myrup, Pitt, Christensen, & Rapoport, 2000). Release of Dox by ultrasound results in a significant improvement on the cellular uptake of drugs and reduction of tumor size in various cancer and tumor models. Real-time quantification by fluorescence studies was employed to measure the amount of Dox released from Pluronic® micelles under continuous or pulse-wave ultrasound in a low frequency range between 20 and 90 kHz. The amount of drug released is most efficient at 20 kHz despite low densities and decreases at higher frequencies despite higher power densities. The transient cavitation threshold increased with increasing frequency, which may indicate the substantial role of transient cavitation on the release. It is worth noting that the amount of drug released under pulse-wave ultrasound is the same as under continuous ultrasound of the same power density. In addition, the released drug was quickly reencapsulated under the pulsed ultrasound suggesting that ultrasound can be used for controlled release of drug from carrier, which can prevent unwanted interaction of drug with normal tissues (Husseini et al., 2000).

Dox uptake from PBS and Pluronic® solution by a HL-60 cell line has been enhanced under both continuous and pulsed ultrasound at 20 kHz. The drug uptake was further enhanced under either higher ultrasound power or longer pulse duration in the range of 0.1–2 s. A threshold time value was found to be at 0.1 s. Under this value, there was no significant change in the amount of drug released even when exposed to 20 kHz ultrasound and with power density of 58 mW/cm^2. The release of hydrophobic dye from Pluronic® micelles triggered by the low-frequency ultrasound also confirms that mechanical cavitation plays a significant role on drug release (Husseini et al., 2000).

Husseini et al. also found that the amount of Dox released from Pluronic® micelles stabilized by interpenetrating network is much lower than that from unstabilized Pluronic® micelles (Husseini & Pitt, 2008). The covalent networks of the stabilized micelles can be disrupted using ultrasound at frequencies 70 and 476 kHz, with a mechanical index of 0.9. Control experiments have also been performed to confirm that the improved accumulation of Dox in HL-60 cells is due to the ultrasound and not due to endocytosis. In addition, significant reduction of tumors in rats that received Dox micelles and with ultrasound triggered at 20 and 70 kHz suggests that ultrasound causes the release of Dox and extravasation of drugs into the tumor cells (Rapoport et al., 1999).

Another study shows that the drug can also be released from pH-responsive PEGylated micelles when triggered by HIFU. Up to 85% of the drug was released from the core of the cross-linked polymer, and the polymer carrier had no observable degradation. This implies that when polymeric micelles are exposed to HIFU, they are temporarily destabilized by the induced shear forces and not due to erosion of the polymer carrier. In a different but similar study, it was shown that the release behavior of Nile red from diblock copolymer, poly(ethylene glycol)-block-poly(lactic acid) (PEG-b-PLA) micelles, can be tuned by adjusting the exposure time to HIFU, ultrasound intensity, and its location (Zhang, Xia, Wang, & Li, 2009). For this system, the release mechanism was based on the scission of the ester bond connecting the PEG and PLA blocks of the micelle structure resulting from transient cavitation in the HIFU focal spot. The breakage of the bond causes an imbalance on hydrophobicity, disruption of the micellar structure, and irreversible release of the drug.

Similarly, Li et al. encapsulated hydrophobic pyrene into micelles comprised of PEG-S-S-PLA, which is PEG-b-PLA with a redox-sensitive disulfide bond as its central linkage. Upon exposure to HIFU, the disulfide bond breaks as confirmed by the sudden decrease of average molecular weight of the block copolymer (Li, Tong, Xia, Zhang, & Xuan, 2010). On the other hand, Tong et al. concluded that micelles composed of polymers with both ester and disulfide bonds (e.g., PEG-COO-SS-PPG) will cleave preferentially at the central ester bond rather than at the disulfide bond (Zhang et al., 2013). Aside from ester and disulfide bonds, the linkage can also be a metal–ligand bond (also known as metallosupramolecular copolymer). In a study, it was shown that 75% of pyrene encapsulated in PPG-[Cu]-PEG micelle can be released at a low HIFU power (Liang, Tong, Wang, Guo, & Xia, 2014).

Ultrasound can also be used to vaporize liquid droplets of perfluorocarbon inside an emulsion-liposome, or eLiposome (Lattin, Belnap, & Pitt, 2012). This technique has been utilized to deliver calcein, plasmids, and Dox to cells by applying ultrasound to vaporize perfluorocarbon to the point that it bursts the micelles and release the drug. Ultrasound can also be applied in combination with gene-carrying liposomes used in genetic targeting of tumor for cancer therapy. DNA can stably associate with cationic liposomes until it is acoustically released (Lattin, Belnap, et al., 2012).

Drug-loaded small liposomes (about 100 nm in diameter) can be attached to larger gas bubbles (1–3 μm in diameter). During insonation, larger bubbles can be disrupted by shear forces and shock waves caused by cavitation, and subsequently, they release the smaller liposomes with drugs. The attachment between liposomes and bubbles can be via biotin–streptavidin, biotin–avidin, or maleimide–thiol (Escoffre et al., 2013; Kheirolomoom et al., 2007; Klibanov et al., 2010). Perfluoropropane bubble (1.6 μm in diameter) with 2.25 MHz ultrasound was successfully used to deliver fluorescent molecules (a mock drug) to human prostate cancer cells. Furthermore, Dox-loaded liposomes were delivered to human glioblastoma cells using 3.6 μm diameter perfluoropropane bubbles and triggered by 1 MHz ultrasound waves. Polymer nanoparticles can also be used instead of liposomes to carry the drug. For example, 5FU, an anticancer agent, was loaded into PLGA nanoparticles, which were then attached to albumin-stabilized MB. Exposure to 1 MHz of ultrasound disrupts the MB and releases the drug, which caused a reduction in tumor growth. Similarly, coinjection of Optison® MB and small quantum dots nanoparticles (20 nm in diameter) were propelled into breast cancer spheroids by cavitation when pulsed ultrasound was applied (Grainger et al., 2011). A list of nanoparticles used for ultrasound-triggered delivery is given in Table 2.

Table 2 Summary of Nanoparticles Used for Ultrasound-Triggered Drug Release

Nanoparticle Type	Application	US Method	Reference
Dox-loaded Pluronic® P-105 micelles	Drug delivery	20–90 kHz	Husseini et al. (2000)
Dox-loaded Pluronic® P-105 micelles stabilized with interpenetrating network of *N,N*-diethylacrylamide	Drug delivery	70–476 kHz	Xu (2011)
Cy5.5-loaded pH-responsive PEGylated micelles	Drug delivery	HIFU	Smith, Eccleston, and Slater (2008)
Nile Red-loaded PLA-b-PEG micelle	Drug delivery	HIFU	Zhang et al. (2009)
Pyrene-loaded PEG-S-S-PLA micelles	Drug delivery	HIFU	Li et al. (2010)
Pyrene-loaded PEG-COO-SS-PPG micelles	Drug delivery	HIFU	Tong, Lu, and Xia (2014)
Pyrene-loaded PPG-[Cu]-PEG micelle	Drug delivery	HIFU	Liang et al. (2014)

Continued

Table 2 Summary of Nanoparticles Used for Ultrasound-Triggered Drug Release—cont'd

Nanoparticle Type	Application	US Method	Reference
Emulsion-liposome, or eLiposome	Vaporize liquid droplets of perfluorohexane (PFC6)	Sonication at 20 kHz	Lattin, Belnap, et al. (2012)
Calcein-loaded PFC5 eLiposomes	Drug delivery	20 kHz ultrasound	Lattin, Pitt, Belnap, and Husseini (2012)
Dox-loaded PFC5 eLiposomes	Drug delivery	20 kHz, 1.0 MHz, and 3.0 MHz	Lin, Javadi, Belnap, Barrow, and Pitt (2014)
Cationic liposomes/ plasmid-DNA complexes (lipoplexes)	Enhance liposome-mediated gene transfer to eukaryotic cells in culture	2 MHz pulsed Doppler US	Koch, Pohl, Cobet, and Rainov (2000)
Biotinylated decafluorobutane microbubbles coated with calcein- and thrombin-loaded biotinylated liposomes via a streptavidin linker	Drug delivery	1 MHz ultrasound	Klibanov et al. (2010)
Microbubbles conjugated to liposomes via avidin and biotin	Drug delivery	2.25 MHz ultrasound	Kheirolomoom et al. (2007)
5FU-bearing nanoparticles bonded to microbubbles	Drug delivery	1 MHz ultrasound	Burke, Alexander, Timbie, Kilbanov, and Price (2014)

5. SIMULTANEOUS IMAGING AND TREATMENT OF CANCER BY ULTRASOUND

In recent years, development of agents that enable simultaneous real-time diagnostic (imaging) and therapeutic (drug delivery), also known as theranostic agents, has gained increasing attention. The use of theranostic agents causes a drastic increase in therapeutic efficiency by allowing the researcher or physician to monitor the accumulation of drugs at the active site (e.g., cancer/tumor) and the timing to trigger drug release at the target site (Allen & Allen, 2014). As previously discussed, ultrasound is widely used due to its low cost, lack of ionizing radiation, ability to be applied to real-time imaging, ease of use, and high spatial and temporal resolution. In addition, ultrasound can be used to control drug release via thermal and cavitation effects at the target sites, which offers more advantages over other traditional stimuli, including pH changes, temperature, and enzyme degradation (Sirsi & Borden, 2014).

Previously, UCAs (specifically MB) have been coinjected with therapeutic agents and were allowed to circulate in the body. Ultrasound imaging of MB enables the visualization of the microvasculature of tumors. As mentioned earlier, MB under ultrasound field result in formation of cavitation that increases the permeability of the cellular membrane and allows the drug to penetrate the intracellular space (Bekeredjian, Katus, Kuecherer, & Bekeredjian, 2006; Kheirolomoom et al., 2007; Kinoshita & Hynynen, 2005). Although this has been shown to improve drug uptake of tumors, a number of side effects have still been observed due to direct injection of drug into the blood stream. The current clinical use of ultrasound for theranostic applications is based on the dual-use of MB as an imaging agent and as a carrier of therapeutic agents. Aside from direct physical loading of the drug into MB, the outer membrane of MB can also have reactive functional groups that can covalently bind the drug and be cleaved by ultrasound.

To further improve the theranostic efficiency of ultrasound, targeting agents, such as protein, peptides, antibody, polymer, and aptamer, are conjugated with a reactive moiety of the shell of the UCA (Nakatsuka, Mattrey, Esener, Cha, & Goodwin, 2012). Conjugation can be achieved in two different ways: covalent coupling (chemical modification) and noncovalent coupling (physical modification). Attaching the ligand of interest to the active site of the shell through biotin–avidin interaction is an example of noncovalent coupling. However, such conjugation strategy has clinical limitations because of its immunogenic nature. Covalent coupling, on the other hand, utilizes carboxylate, maleimide, or thiol groups as reactive functional groups and is a more common modification technique (Nakatsuka et al., 2012). The covalent coupling is accomplished via direct chemical reaction, which does not need intermediate molecules such as streptavidin so the chance of recognition by the immune system is significantly lower (Nakatsuka et al., 2012).

For cancer therapy, the targeting agents decorated on the shell of UCA are typically ligand-mediated to target specific biomarkers in tumors. Angiogenesis biomarkers that are commonly targeted include vascular endothelial growth factor (VEGF), vascular endothelial growth factor receptor 2 (VEGFR2), endoglin (CD105), and $\alpha_v\beta_3$ integrin (Alzaraa et al., 2012; Fokong et al., 2013; Pochon et al., 2010). On the other hand, one of the most common utilized targeting agents is the clinically approved cyclic RGD (cRGD). UCA labeled with cRGD has been shown to have increased residence time because it binds to the endothelial cells expressing $\alpha_v\beta_3$ integrin (Jun, Park, Kim, & Yoon, 2010).

Although microsized UCAs have wide clinical applications, the relatively large size limits its applications in imaging and delivery to the vasculature, and its somewhat short in vivo half-life results in insufficient imaging frames of the target. Consequently, recent attention has been focused on the development of submicron- or nanosized UCAs (e.g., NB) that are capable of accumulating and extravasating in cancer tissues due to the enhanced permeability and retention effect (EPR). The nanoscale particles permit satisfactory imaging and on-site drug release for a theranostic platform.

A number of theranostic UCAs are formulated by loading UCA with supermagnetic iron oxide nanoparticles (SPIONs), CuS nanoparticles, gold nanoparticles (GNP), and gold rods (GNR), among others. For instance, Wang et al. loaded UCA with GNR by sonication of solution containing GNR, human serum albumin, and avidin with C_3F_8 as the headspace gas. The avidin on the UCA was then modified with VEGFR2 attached to biotin. Ultrasound was used to image and identify angiogenesis region in the tumor, while near-infrared radiation (NIR) induced the photoacoustic imaging and photothermal therapy by GNR.

Lipid-based nanocarriers or liposomes have additional advantages such as controllable size, modifiable shell for active targeting, and ability to load both hydrophilic and hydrophobic drugs for therapeutic effect. Liposomes that respond to ultrasound, i.e., echogenic liposome (ELIPs), are formulated by introducing gas pockets inside the liposomes. ELIPs can be used to encapsulate drugs, antigens, or genes, and have them release at the target by an ultrasound trigger (Guo & Szoka, 2003).

Polymer-based UCAs have also gained increased attention due to their biocompatibility and biodegradability. In this approach, the polymeric UCAs, typically composed of PLA, deliver the drug as a response to an external focused ultrasound (Guo & Szoka, 2003). Ke et al. designed a PLA theranostic platform by loading GNR in PLA microcapsules for combined ultrasound contrast imaging and photothermal therapy (Guo & Szoka, 2003).

In a related study, PLGA NB were synthesized using double emulsion evaporation techniques and demonstrated an on-demand drug release with HIFU and enhanced therapeutic efficacy with an improved therapeutic index compared to systemically administered chemotherapeutic agents (Rapoport et al., 2011). Rapoport et al. established a new generation of perfluorocarbon nanoemulsion by introducing perfluoro-15-crown-5-ether (PFCE). The PFCE nanodroplets demonstrated both ultrasound and fluorine (19F) MR imaging properties and provided cargo for therapeutic drugs (paclitaxel, PTX). The PTX-loaded droplets showed significantly greater inhibition of tumor growth as opposed to empty droplets (Fig. 5) (Rapoport et al., 2011; Ta et al., 2014).

Shuail et al. developed siRNA-loaded nanoscale contrast agents by combining siRNA polymeric micelles and gas-cored lipid NB (siRNA-NB). The high tumor accumulation of siRNA-NB was achieved due to its small size, which allowed effective delivery into the tumor cells with the aid of ultrasound for image-guided gene therapy (Rapoport et al., 2011).

Another approach is to use nanosized UCA composed of liquid PFCs such as perfluorooctyl bromide (PFOB) encapsulated by a polymer shell. Due to its small size, such UCA shows improved accumulation in tissues resulting in a satisfactory image-guided therapy (Kheirolomoom et al., 2007; Lanza et al., 1997). A dual-functional theranostic platform was also achieved by incorporating a gold nanoshell into a PFOB nanocapsule and using ultrasound/CT image-guided photothermal ablation of the tumor (Ke et al., 2014; Ta et al., 2014).

In recent studies, Lee et al. designed an efficient nanosized theranostic agent that can generate gas at the active site. Calcium carbonate was encapsulated in PLGA nanoparticles, which when exposed to acidic environment in the tumor generates carbon dioxide allowing contrast-enhanced real-time US imaging. It has also been observed that the generation of intracellular gas bubbles induced necrotic cell death (Lee et al., 2016).

As an alternate approach, Exner et al. developed monolayered, lipid-shelled, nanosized UCA stabilized by Pluronic® copolymer as shown in Fig. 6 (Perera et al., 2015, 2014). Integration of Pluronic® into the NB provides stability of the shell structure, reduction of the size of the particles, and sensitization for low-grade hyperthermia (Perera et al., 2013). Further modification of the membrane of the UCA by incorporating interpenetrating and biodegradable cross-linking agent increases its circulation time in the blood stream and permit to load therapeutic drug for theranostic applications for ovarian tumor (Nittayacharn, Lilly, & Exner, 2017). In addition, Jiang et al. conjugated Herceptin, HER-2-specific antibody-based anticancer

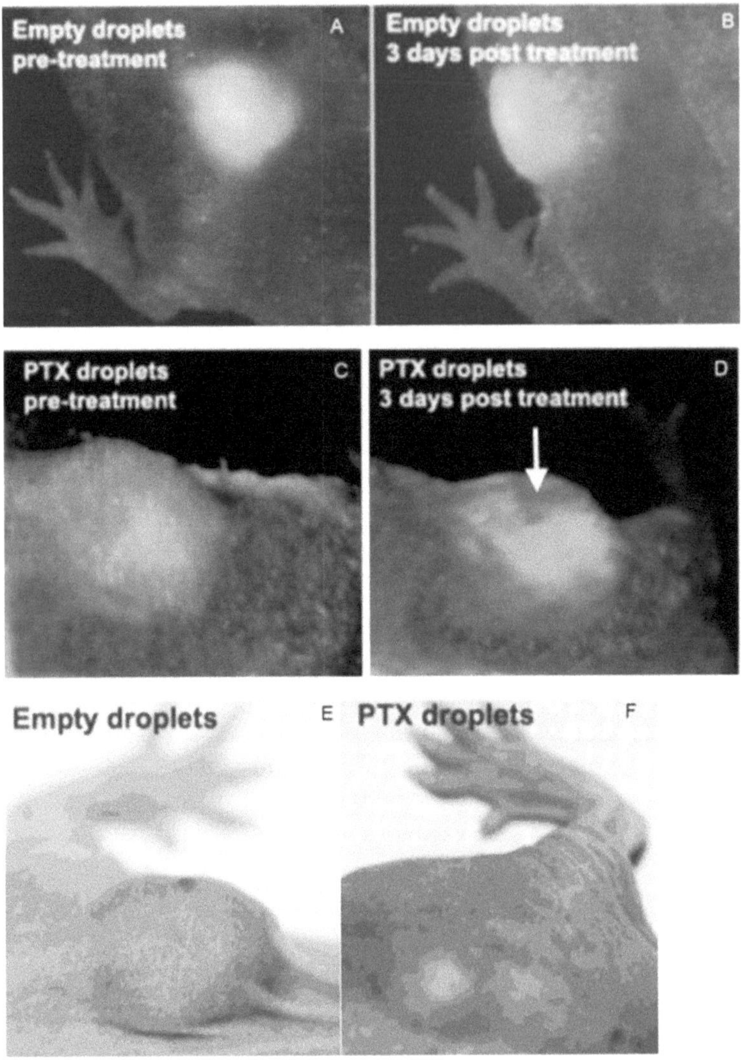

Fig. 5 Intravital fluorescence images of subcutaneous pancreatic tumors before and after FUS treatment (A–D). Mice were injected with empty droplets (A, B) or PTX-loaded droplets (C, D) (Rapoport et al., 2011). *Photographs of the tumors 12 days after treatment for empty droplets (E) and PTX-loaded droplets (F).*

agents that target breast cancer, to the NB with the same formulation (Jiang et al., 2016). Furthermore, targeting NB with prostate-specific membrane antigen (PSMA) for different prostate cancer models for image-guided prostate cancer treatments has gained enormous attention (Fan et al., 2015). A summary of the theranostic nanocarriers discussed is included in Table 3.

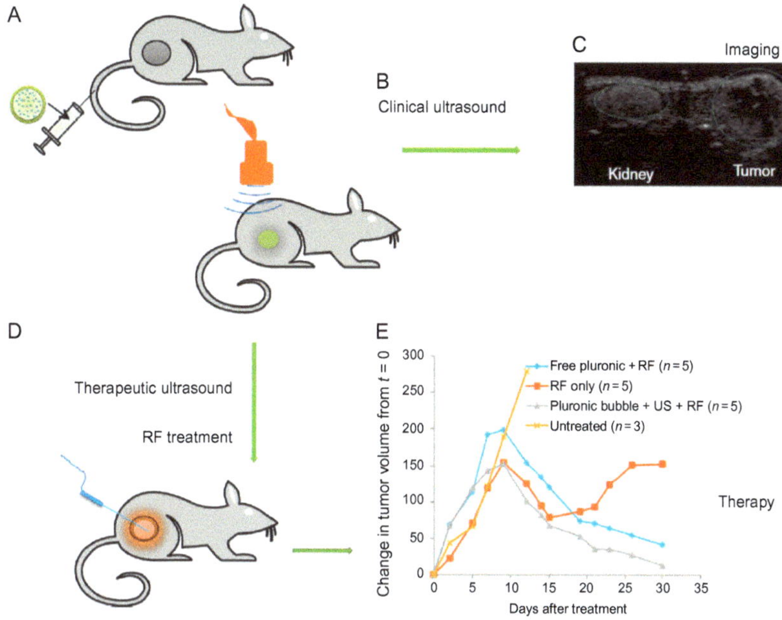

Fig. 6 Theranostic effect of Pluronic nanobubbles showing the injection (A), ultrasound imaging (B), representative contrast-enhanced images of tumor and kidney after bubble injection (C), and the use of ultrasound for tumor treatment (D, E) (Perera et al., 2015).

Table 3 Summary of Theranostic Nanocarriers With Their Therapeutic Agent and Targeting Molecule

Nanocarrier	Material	Gas	Targeting	Therapeutic Agent	References
Gold nanoshell	Gold particles	C_3F_8	VEGFR2	Gold nanorods	Ke et al. (2014)
Echogenic liposome	Lipid bilayer	C_3F_8	NA	Drugs and genes	Schroeder, Kost, and Barenholz (2009), Suzuki et al. (2007)
Triblock (Pluronic) NB	Lipid monolayer	C_3F_8	Cancer Antigen-125	NA	Gao et al. (2017), Perera et al. (2011, 2014)
Micelle/ polymeric nanocarriers	PEG/PLGA/ PAA	NA	NA	Drug-Methotrexate, Doxorubicin	Rapoport et al. (2011)
Nanodroplets	Lipids	C_3F_8	NA	NA	Ta et al. (2014)

6. CONCLUSION

Ultrasound is a flexible platform for affordable, robust, and real-time imaging. With UCAs used as a drug carrier, ultrasound can also be used as a fully noninvasive, on-demand trigger mechanism to induce thermal or cavitation effect to release the encapsulated drug and improved influx of drugs through the formation of transient openings on the cell membrane. With proper adaptation, the techniques are broadly applicable to commonly utilized drug delivery carriers including micelles, liposomes, and solid nanoparticles and can play a key role in increasing drug distribution and cell uptake in a region of interests, thereby boosting therapeutic efficacy. Overall ultrasound offers an exciting and timely opportunity to enhance nanomedicine-based delivery technologies and have significant impact in future cancer therapeutic development.

REFERENCES

Agarwal, A., Ng, W. J., & Liu, Y. (2011). Principle and applications of microbubble and nanobubble technology for water treatment. *Chemosphere*, *84*(9), 1175–1180. https://doi.org/10.1016/j.chemosphere.2011.05.054.

Aggeli, C., Giannopoulos, G., Roussakis, G., Christoforatou, E., Marinos, G., Toli, C., et al. (2008). Safety of myocardial flash-contrast echocardiography in combination with dobutamine stress testing for the detection of ischaemia in 5250 studies. *Heart*, *94*(12), 1571–1577. https://doi.org/10.1136/hrt.2007.135145.

Allen, T. M., & Allen, T. M. (2014). Drug delivery systems: Entering the mainstream. *Science*, *303*, 1818–1823. https://doi.org/10.1126/science.1095833.

Alzaraa, A., Gravante, G., Chung, W. Y., Al-Leswas, D., Bruno, M., Dennison, A. R., et al. (2012). Targeted microbubbles in the experimental and clinical setting. *American Journal of Surgery*, *204*(3), 355–366. https://doi.org/10.1016/j.amjsurg.2011.10.024.

Apfel, R. E., & Holland, C. K. (1991). Gauging the likelihood of cavitation from short-pulse, low-duty cycle diagnostic ultrasound. *Ultrasound in Medicine and Biology*, *17*(2), 179–185. https://doi.org/10.1016/0301-5629(91)90125-G.

Bader, K. B., & Holland, C. K. (2013). Gauging the likelihood of stable cavitation from ultrasound contrast agents. *Physics in Medicine & Biology*, *58*(1), 127–144. https://doi.org/10.1088/0031-9155/58/1/127.

Baker, K. G., Robertson, V. J., & Duck, F. A. (2001). A review of therapeutic ultrasound: Effectiveness studies. *Physical Therapy*, *81*(7), 1351–1358. https://doi.org/10.1093/ptj/81.7.1339.

Barnett, S. B., ter Haar, G. R., Ziskin, M. C., Nyborg, W. L., Maeda, K., & Bang, J. (1994). Current status of research on biophysical effects of ultrasound. Ultrasound in Medicine & Biology, 20(3), 205–218. Retrieved from: http://www.ncbi.nlm.nih.gov/pubmed/8059482

Bekeredjian, R., Katus, H. A., Kuecherer, H. F., & Bekeredjian, R. (2006). Therapeutic use of ultrasound targeted microbubble destruction: A review of non-cardiac applications. *Ultraschall in der Medizin*, *27*(2), 134–140. https://doi.org/10.1055/s-2005-858993.

Burgess, A., Shah, K., Hough, O., & Hynynen, K. (2015). Focused ultrasound-mediated drug delivery through the blood-brain barrier. *Expert Review of Neurotherapeutics, 15*(5), 477–491. https://doi.org/10.1586/14737175.2015.1028369.

Burke, C. W., Alexander, E., Timbie, K., Kilbanov, A. L., & Price, R. J. (2014). Ultrasound-activated agents comprised of 5FU-bearing nanoparticles bonded to microbubbles inhibit solid tumor growth and improve survival. *Molecular Therapy: The Journal of the American Society of Gene Therapy, 22*(2), 321–328. https://doi.org/10.1038/mt.2013.259.

Cai, W. B., Yang, H. L., Zhang, J., Yin, J. K., Yang, Y. L., Yuan, L. J., et al. (2015). The optimized fabrication of nanobubbles as ultrasound contrast agents for tumor imaging. *Scientific Reports, 5*, 1–11. https://doi.org/10.1038/srep13725.

Calliada, F., Campani, R., Bottinelli, O., Bozzini, A., & Sommaruga, M. G. (1998). Ultrasound contrast agents: Basic principles. *European Journal of Radiology, 27*(Suppl. 2), 157–160. https://doi.org/10.1016/S0720-048X(98)00057-6.

Chemat, F., Rombaut, N., Sicaire, A. G., Meullemiestre, A., Fabiano-Tixier, A. S., & Abert-Vian, M. (2017). Ultrasound assisted extraction of food and natural products. Mechanisms, techniques, combinations, protocols and applications. A review. *Ultrasonics Sonochemistry, 34*, 540–560. https://doi.org/10.1016/j.ultsonch.2016.06.035.

Couture, O., Foley, J., Kassell, N. F., Larrat, B., & Aubry, J.-F. (2014). Review of ultrasound mediated drug delivery for cancer treatment: Updates from pre-clinical studies. *Translational Cancer Research, 3*(5), 494–511. https://doi.org/10.3978/j.issn.2218-676X.2014.10.01.

Dawson, P. (2002). Ultrasound contrast agents. *European Journal of Radiology, 41*(3), 175. https://doi.org/10.1016/S0720-048X(02)00022-0.

Dejong, N., Tencate, F. J., Lancee, C. T., Roelandt, J. R. T. C., & Bom, N. (1991). Principles and recent developments in ultrasound contrast agents. *Ultrasonics, 29*(4), 324–330. https://doi.org/10.1016/0041-624x(91)90030-C.

Dolan, M. S., Gala, S. S., Dodla, S., Abdelmoneim, S. S., Xie, F., Cloutier, D., et al. (2009). Safety and efficacy of commercially available ultrasound contrast agents for rest and stress echocardiography. A multicenter experience. *Journal of the American College of Cardiology, 53*(1), 32–38. https://doi.org/10.1016/j.jacc.2008.08.066.

Escoffre, J. M., Mannaris, C., Geers, B., Novell, A., Lentacker, I., Averkiou, M., et al. (2013). Doxorubicin liposome-loaded microbubbles for contrast imaging and ultrasound-triggered drug delivery. *IEEE Transactions on Ultrasonics, Ferroelectrics, and Frequency Control, 60*(1), 78–87. https://doi.org/10.1109/TUFFC.2013.2539.

Fan, X., Wang, L., Guo, Y., Tu, Z., Li, L., Tong, H., et al. (2015). Ultrasonic nanobubbles carrying anti-PSMA nanobody: Construction and application in prostate cancer-targeted imaging. *PLoS One, 10*(6), 1–13. https://doi.org/10.1371/journal.pone.0127419.

Feinstein, S. B., Cheirif, J., Ten Cate, F. J., Silverman, P. R., Heidenreich, P. A., Dick, C., et al. (1990). Safety and efficacy of a new transpulmonary ultrasound contrast agent: Initial multicenter clinical results. *Journal of the American College of Cardiology, 16*(2), 316–324. https://doi.org/10.1016/0735-1097(90)90580-I.

Fenster, A., & Downey, D. B. (1996). 3-D ultrasound imaging: A review. *IEEE Engineering in Medicine and Biology Magazine, 15*, 41–51. https://doi.org/10.1109/51.544511.

Fokong, S., Fragoso, A., Rix, A., Curaj, A., Wu, Z., Lederle, W., et al. (2013). Ultrasound molecular imaging of E-Selectin in tumor vessels using poly n-butyl cyanoacrylate microbubbles covalently coupled to a short targeting peptide. *Investigative Radiology, 48*(12), 843–850. https://doi.org/10.1097/RLI.0b013e31829d03ec.

Frangioni, J. V. (2008). New technologies for human cancer imaging. *Journal of Clinical Oncology, 26*(24), 4012–4021. https://doi.org/10.1200/JCO.2007.14.3065.

Gao, Y., Hernandez, C., Yuan, H.-X., Lilly, J., Kota, P., Zhou, H., et al. (2017). Ultrasound molecular imaging of ovarian cancer with CA-125 targeted nanobubble contrast agents.

Nanomedicine: Nanotechnology, Biology, and Medicine, 13(7), 2159–2168. https://doi.org/10.1016/j.nano.2017.06.001.

Goldberg, B. B., Liu, J. B., & Forsberg, F. (1994). Ultrasound contrast agents: A review. *Ultrasound in Medicine & Biology, 20*(4), 319–333. https://doi.org/10.1016/0301-5629(94)90001-9.

Grainger, S. J., Serna, J. V., Sunny, S., Zhou, Y., Deng, C. X., & El-sayed, M. E. H. (2011). Pulsed ultrasound enhance nanoparticle penetration into breast cancer spheroids. *Molecular Pharmaceutics*, 7(6), 2006–2019. https://doi.org/10.1021/mp100280bPulsed.

Guo, X., & Szoka, F. C. (2003). Chemical approaches to triggerable lipid vesicles for drug and gene delivery. *Accounts of Chemical Research, 36*(5), 335–341. https://doi.org/10.1021/ar9703241.

Hernandez, C., Gulati, S., Fioravanti, G., Stewart, P. L., & Exner, A. A. (2017). Cryo-EM visualization of lipid and polymer-stabilized perfluorocarbon gas nanobubbles—A step towards nanobubble mediated drug delivery. *Scientific Reports*, 7(1), 1–8. https://doi.org/10.1038/s41598-017-13741-1.

Husseini, G. A., Myrup, G. D., Pitt, W. G., Christensen, D. A., & Rapoport, N. Y. (2000). Factors affecting acoustically triggered release of drugs from polymeric micelles. *Journal of Controlled Release, 69*, 43–52. https://doi.org/10.1016/S0168-3659(00)00278-9.

Husseini, G. A., & Pitt, W. G. (2008). Micelles and nanoparticles for ultrasonic drug and gene delivery. *Advanced Drug Delivery Reviews, 60*(10), 1137–1152. https://doi.org/10.1016/j.addr.2008.03.008.

Hynynen, K. (2010). MRI-guided focused ultrasound treatments. *Ultrasonics, 50*(2), 221–229. https://doi.org/10.1016/j.ultras.2009.08.015.

Ignee, A., Atkinson, N. S. S., Schuessler, G., & Dietrich, C. F. (2016). Ultrasound contrast agents. *Endoscopic Ultrasound, 5*(6), 355–362. https://doi.org/10.4103/2303-9027.193594.

Jiang, Q., Hao, S., Xiao, X., Yao, J., Ou, B., Zhao, Z., et al. (2016). Production and characterization of a novel long-acting Herceptin-targeted nanobubble contrast agent specific for Her-2-positive breast cancers. *Breast Cancer, 23*(3), 445–455. https://doi.org/10.1007/s12282-014-0581-8.

Jun, H. Y., Park, S. H., Kim, H. S., & Yoon, K. H. (2010). Long residence time of ultrasound microbubbles targeted to integrin in murine tumor model. *Academic Radiology, 17*(1), 54–60. https://doi.org/10.1016/j.acra.2009.07.017.

Ke, H., Yue, X., Wang, J., Xing, S., Zhang, Q., Dai, Z., et al. (2014). Gold nanoshelled liquid perfluorocarbon nanocapsules for combined dual modal ultrasound/ct imaging and photothermal therapy of cancer. *Small, 10*(6), 1220–1227. https://doi.org/10.1002/smll.201302252.

Kheirolomoom, A., Dayton, P. A., Lum, A. F. H., Little, E., Paoli, E. E., Zheng, H., et al. (2007). Acoustically-active microbubbles conjugated to liposomes: Characterization of a proposed drug delivery vehicle. *Journal of Controlled Release, 118*(3), 275–284. https://doi.org/10.1016/j.jconrel.2006.12.015.

Kinoshita, M., & Hynynen, K. (2005). Intracellular delivery of Bak BH3 peptide by microbubble-enhanced ultrasound. *Pharmaceutical Research, 22*(5), 716–720. https://doi.org/10.1007/s11095-005-2586-7.

Klibanov, A. L., Shevchenko, T. I., Raju, B. I., Seip, R., & Chin, C. T. (2010). Ultrasound-triggered release of materials entrapped in microbubble-liposome constructs: A tool for targeted drug delivery. *Journal of Controlled Release, 148*(1), 13–17. https://doi.org/10.1016/j.jconrel.2010.07.115.

Koch, S., Pohl, P., Cobet, U., & Rainov, N. G. (2000). Ultrasound enhancement of liposome-mediated cell transfection is caused by cavitation effects. *Ultrasound in Medicine and Biology, 26*(5), 897–903. https://doi.org/10.1016/S0301-5629(00)00200-3.

Kost, J., Leong, K., & Langer, R. (1989). Ultrasound-enhanced polymer degradation and release of incorporated substances. *Proceedings of the National Academy of Sciences of the United States of America, 86*(20), 7663–7666. https://doi.org/10.1073/pnas.86.20.7663.

Kovacs, Z. I., Kim, S., Jikaria, N., Qureshi, F., Milo, B., Lewis, B. K., et al. (2017). Disrupting the blood-brain barrier by focused ultrasound induces sterile inflammation. *Proceedings of the National Academy of Sciences of the United States of America, 114*(1), E75–E84. https://doi.org/10.1073/pnas.1614777114.

Lanza, G. M., Wallace, K. D., Fischer, S. E., Christy, D. H., Scott, M. J., Trousil, R. L., et al. (1997). High-frequency ultrasonic detection of thrombi with a targeted contrast system. *Ultrasound in Medicine and Biology, 23*(6), 863–870. https://doi.org/10.1016/S0301-5629(97)00046-X.

Lattin, J. R., Belnap, D. M., & Pitt, W. G. (2012). Formation of eLiposomes as a drug delivery vehicle. *Colloids and Surfaces B: Biointerfaces, 89*(1), 93–100. https://doi.org/10.1016/j.colsurfb.2011.08.030.

Lattin, J. R., Pitt, W. G., Belnap, D. M., & Husseini, G. A. (2012). Ultrasound-induced calcein release from eLiposomes. *Ultrasound in Medicine and Biology, 38*(12), 2163–2173. https://doi.org/10.1016/j.ultrasmedbio.2012.08.001.

Lee, M. H., Lin, H. Y., Chen, H. C., & Thomas, J. L. (2008). Ultrasound mediates the release of curcumin from microemulsions. *Langmuir, 24*(5), 1707–1713. https://doi.org/10.1021/la7022874.

Lee, J., Min, H. S., You, D. G., Kim, K., Kwon, I. C., Rhim, T., et al. (2016). Theranostic gas-generating nanoparticles for targeted ultrasound imaging and treatment of neuroblastoma. *Journal of Controlled Release, 223*, 197–206. https://doi.org/10.1016/j.jconrel.2015.12.051.

Lentacker, I., De Smedt, S. C., & Sanders, N. N. (2009). Drug loaded microbubble design for ultrasound triggered delivery. *Soft Matter, 5*(11), 2161. https://doi.org/10.1039/b823051j.

Li, Y., Tong, R., Xia, H., Zhang, H., & Xuan, J. (2010). High intensity focused ultrasound and redox dual responsive polymer micelles. *Chemical Communications, 46*(41), 7739. https://doi.org/10.1039/c0cc02628j.

Liang, B., Tong, R., Wang, Z., Guo, S., & Xia, H. (2014). High intensity focused ultrasound responsive metallo-supramolecular block copolymer micelles. *Langmuir, 30*(31), 9524–9532. https://doi.org/10.1021/la500841x.

Lin, C.-Y., Javadi, M., Belnap, D. M., Barrow, J. R., & Pitt, W. G. (2014). Ultrasound sensitive eLiposomes containing doxorubicin for drug targeting therapy. *Nanomedicine: Nanotechnology, Biology, and Medicine, 10*(1), 67–76. https://doi.org/10.1016/j.nano.2013.06.011.

Lin, H. Y., & Thomas, J. L. (2003). PEG-lipids and oligo (ethylene glycol) surfactants enhance the ultrasonic permeabilizability of liposomes. *Langmuir, 19*(4), 1098–1105. https://doi.org/10.1021/la026604t.

Lindner, J. R. (2004). Microbubbles in medical imaging: Current applications and future directions. *Nature Reviews Drug Discovery, 3*(6), 527–533. https://doi.org/10.1038/nrd1417.

Magara, A., Bühler, R., Moser, D., Kowalski, M., Pourtehrani, P., & Jeanmonod, D. (2014). First experience with MR-guided focused ultrasound in the treatment of Parkinson's disease. *Journal of Therapeutic Ultrasound, 2*(1), 11. https://doi.org/10.1186/2050-5736-2-11.

Main, M. L., Ryan, A. C., Davis, T. E., Albano, M. P., Kusnetzky, L. L., & Hibberd, M. (2008). Acute mortality in hospitalized patients undergoing echocardiography with and without an ultrasound contrast agent (multicenter registry results in 4,300,966 consecutive patients). *American Journal of Cardiology, 102*(12), 1742–1746. https://doi.org/10.1016/j.amjcard.2008.08.019.

Mehier-Humbert, S., Bettinger, T., Yan, F., & Guy, R. H. (2005). Plasma membrane poration induced by ultrasound exposure: Implication for drug delivery. *Journal of Controlled Release, 104*(1), 213–222. https://doi.org/10.1016/j.jconrel.2005.01.007.

Mitragotri, S. (2005). Healing sound: The use of ultrasound in drug delivery and other therapeutic applications. *Nature Reviews Drug Discovery, 4*(3), 255–260. https://doi.org/10.1038/nrd1662.

Nakatsuka, M. A., Mattrey, R. F., Esener, S. C., Cha, J. N., & Goodwin, A. P. (2012). Aptamer-crosslinked microbubbles: Smart contrast agents for thrombin-activated ultrasound imaging. *Advanced Materials, 24*(45), 6010–6016. https://doi.org/10.1002/adma.201201484.

Nittayacharn, P., Lilly, J., & Exner, A. (2017). In *Maximizing drug loading in microbubbles for ultrasound-mediated drug delivery. Poster session presented at the meeting of the Biomedical Engineering Society Annual Meeting, Phoenix, Arizona.*

Oerlemans, C., Bult, W., Bos, M., Storm, G., Nijsen, J. F. W., & Hennink, W. E. (2010). Polymeric micelles in anticancer therapy: Targeting, imaging and triggered release. *Pharmaceutical Research, 27*(12), 2569–2589. https://doi.org/10.1007/s11095-010-0233-4.

Oláh, L. (2016). In C. B. László Csiba (Ed.), *Ultrasound principles.* Cambridge University Press.

Paefgen, V., Doleschel, D., & Kiessling, F. (2015). Evolution of contrast agents for ultrasound imaging and ultrasound-mediated drug delivery. *Frontiers in Pharmacology, 6,* 1–16. https://doi.org/10.3389/fphar.2015.00197.

Park, K. (2010). Focused ultrasound for targeted nanoparticle delivery to tumors. *Journal of Controlled Release, 146*(3), 263. https://doi.org/10.1016/j.jconrel.2010.08.001.

Perera, R. H., Hernandez, C., Zhou, H., Kota, P., Burke, A., & Exner, A. A. (2015). Ultrasound imaging beyond the vasculature with new generation contrast agents. *Wiley Interdisciplinary Reviews: Nanomedicine and Nanobiotechnology,* 7(4), 593–608. https://doi.org/10.1002/wnan.1326.

Perera, R. H., Krupka, T. M., Wu, H., Traughber, B., Dremann, D., Broome, A.-M., et al. (2011). Role of Pluronic block copolymers in modulation of heat shock protein 70 expression. *International Journal of Hyperthermia: The Official Journal of European Society for Hyperthermic Oncology, North American Hyperthermia Group,* 27(7), 672–681. https://doi.org/10.3109/02656736.2011.608218.

Perera, R. H., Patel, R., Wu, H., Gangolli, M., Traughber, B., Oleinick, N., et al. (2013). Preclinical evaluation of radiosensitizing activity of Pluronic block copolymers. *International Journal of Radiation Biology, 89*(10), 801–812. https://doi.org/10.3109/09553002.2013.800246.

Perera, R. H., Solorio, L., Wu, H., Gangolli, M., Silverman, E., Hernandez, C., et al. (2014). Nanobubble ultrasound contrast agents for enhanced delivery of thermal sensitizer to tumors undergoing radiofrequency ablation. *Pharmaceutical Research, 31*(6), 1407–1417. https://doi.org/10.1007/s11095-013-1100-x.

Phenix, C. P., Togtema, M., Pichardo, S., Zehbe, I., & Curiel, L. (2014). High intensity focused ultrasound technology, applications in therapy and drug delivery its scope and applications in therapy and drug delivery. *Journal of Pharmacy & Pharmaceutical Sciences: A Publication of the Canadian Society for Pharmaceutical Sciences, Societe Canadienne Des Sciences Pharmaceutiques,* 17(1), 136–153. https://doi.org/10.18433/J3ZP5F.

Pitt, W. G., Husseini, G. A., & Staples, B. J. (2004). Ultrasonic drug delivery–A general review. *Expert Opinion on Drug Delivery, 1*(1), 37–56. https://doi.org/10.1517/17425247.1.1.37.

Pochon, S., Tardy, I., Bussat, P., Bettinger, T., Brochot, J., Von Wronski, M., et al. (2010). BR55: A lipopeptide-based VEGFR2-targeted ultrasound contrast agent for molecular imaging of angiogenesis. *Investigative Radiology, 45*(2), 89–95. https://doi.org/10.1097/RLI.0b013e3181c5927c.

Qin, S., Caskey, C. F., & Ferrara, K. W. (2009). Ultrasound contrast microbubbles in imaging and therapy: Physical principles and engineering. *Physics in Medicine and Biology, 54*(6) https://doi.org/10.1088/0031-9155/54/6/R01.

Rapoport, N. Y., Herron, J. N., Pitt, W. G., & Pitina, L. (1999). Micellar delivery of doxorubicin and its paramagnetic analog, ruboxyl, to HL-60 cells: Effect of micelle structure and ultrasound on the intracellular drug uptake. *Journal of Controlled Release, 58*(2), 153–162. https://doi.org/10.1016/S0168-3659(98)00149-7.

Rapoport, N., Nam, K. H., Gupta, R., Gao, Z., Mohan, P., Payne, A., et al. (2011). Ultrasound-mediated tumor imaging and nanotherapy using drug loaded, block copolymer stabilized perfluorocarbon nanoemulsions. *Journal of Controlled Release, 153*(1), 4–15. https://doi.org/10.1016/j.jconrel.2011.01.022.

Robertson, V. J., & Baker, K. G. (2001). A review of therapeutic ultrasound: Effectiveness studies. *Physical Therapy, 81*(7), 1339–1350. https://doi.org/10.1093/ptj/81.7.1339.

Sancheti, S. V., & Gogate, P. R. (2017). A review of engineering aspects of intensification of chemical synthesis using ultrasound. *Ultrasonics Sonochemistry, 36*, 527–543. https://doi.org/10.1016/j.ultsonch.2016.08.009.

Sboros, V. (2008). Response of contrast agents to ultrasound. *Advanced Drug Delivery Reviews, 60*(10), 1117–1136. https://doi.org/10.1016/j.addr.2008.03.011.

Schroeder, A., Kost, J., & Barenholz, Y. (2009). Ultrasound, liposomes, and drug delivery: Principles for using ultrasound to control the release of drugs from liposomes. *Chemistry and Physics of Lipids, 162*, 1–2), 1–16. https://doi.org/10.1016/j.chemphyslip.2009.08.003.

Shchukin, D. G., Gorin, D. A., & Möhwald, H. (2006). Ultrasonically induced opening of polyelectrolyte microcontainers. *Langmuir, 22*(17), 7400–7404. https://doi.org/10.1021/la061047m.

Shekhar, H., Smith, N. J., Raymond, J. L., & Holland, C. K. (2018). Effect of temperature on the size distribution, shell properties, and stability of Definity®. *Ultrasound in Medicine & Biology, 44*(2), 434–446.

Sirsi, S. R., & Borden, M. A. (2014). State-of-the-art materials for ultrasound-triggered drug delivery. *Advanced Drug Delivery Reviews, 72*, 3–14. https://doi.org/10.1016/j.addr.2013.12.010.

Sistrom, C. L., & McKay, N. L. (2005). Costs, charges, and revenues for hospital diagnostic imaging procedures: Differences by modality and hospital characteristics. *Journal of the American College of Radiology, 2*(6), 511–519. https://doi.org/10.1016/j.jacr.2004.09.013.

Smith, M. J., Eccleston, M. E., & Slater, N. K. (2008). The effect of high intensity focussed ultrasound (HIFU) on pH responsive PEGylated micelles. *The Journal of the Acoustical Society of America, 123*(5), 3223. https://doi.org/10.1121/1.2933428.

Stouffer, J. R. (2004). History of ultrasound in animal science. *Journal of Ultrasound in Medicine, 23*(5), 577–584. https://doi.org/10.7863/jum.2004.23.5.577.

Suzuki, R., Takizawa, T., Negishi, Y., Hagisawa, K., Tanaka, K., Sawamura, K., et al. (2007). Gene delivery by combination of novel liposomal bubbles with perfluoropropane and ultrasound. *Journal of Controlled Release: Official Journal of the Controlled Release Society, 117*(1), 130–136. https://doi.org/10.1016/j.jconrel.2006.09.008.

Szabo, T. L. (2013). *Diagnostic ultrasound imaging: Inside out*. Academic Press.

Ta, T., Bartolak-Suki, E., Park, E.-J., Karrobi, K., McDannold, N. J., & Porter, T. M. (2014). Localized delivery of doxorubicin in vivo from polymer-modified thermosensitive liposomes with MR-guided focused ultrasound-mediated heating. *Journal of Controlled Release: Official Journal of the Controlled Release Society, 194*, 71–81. https://doi.org/10.1016/j.jconrel.2014.08.013.

Tong, R., Lu, X., & Xia, H. (2014). A facile mechanophore functionalization of an amphiphilic block copolymer towards remote ultrasound and redox dual stimulus responsiveness. *Chemical Communications, 50*(27), 3575. https://doi.org/10.1039/c4cc00103f.

Tsui, B. C. H., & Suresh, S. (2010). Ultrasound imaging for regional anesthesia in infants, children, and adolescents. *Anesthesiology, 112*(2), 473–492. https://doi.org/10.1097/ALN.0b013e3181c5dfd7.

Wrenn, S. P., Dicker, S. M., Small, E. F., Dan, N. R., Mleczko, M., Schmitz, G., et al. (2012). Bursting bubbles and bilayers. *Theranostics, 2*(12), 1140–1159. https://doi.org/10.7150/thno.4305.

Wu, H., Rognin, N. G., Krupka, T. M., Solorio, L., Yoshiara, H., Guenette, G., et al. (2013). Acoustic characterization and pharmacokinetic analyses of new nanobubble ultrasound contrast agents. *Ultrasound in Medicine and Biology*, *39*(11), 2137–2146. https://doi.org/10.1016/j.ultrasmedbio.2013.05.007.

Xu, R. X. (2011). Multifunctional microbubbles and nanobubbles for photoacoustic imaging. *Contrast Media and Molecular Imaging*, *6*(5), 401–411. https://doi.org/10.1002/cmmi.442.

Yasuda, K., Mukai, H., Fujimoto, S., Nakajima, M., & Kawai, K. (1988). The diagnosis of pancreatic cancer by endoscopic ultrasonography. *Gastrointestinal Endoscopy*, *34*(1), 1–8. https://doi.org/10.1016/S0016-5107(88)71220-1.

Young, S. R., & Dyson, M. (1990). Effect of therapeutic ultrasound on the healing of full-thickness excised skin lesions. *Ultrasonics*, *28*(3), 175–180. https://doi.org/10.1016/0041-624X(90)90082-Y.

Yu, A. C. H., & Hossack, J. A. (2017). Methods and protocols: Engine for the next wave of biomedical ultrasound innovations. *IEEE Transactions on Ultrasonics, Ferroelectrics, and Frequency Control*, *64*(1), 6–10. https://doi.org/10.1109/TUFFC.2016.2639698.

Zhang, H., Xia, H., Wang, J., & Li, Y. (2009). High intensity focused ultrasound-responsive release behavior of PLA-b-PEG copolymer micelles. *Journal of Controlled Release*, *139*(1), 31–39. https://doi.org/10.1016/j.jconrel.2009.05.037.

Zhang, Z., Yin, L., Tu, C., Song, Z., Zhang, Y., Xu, Y., et al. (2013). Redox-responsive, core cross-linked polyester micelles. *ACS Macro Letters*, *2*(1), 40–44. https://doi.org/10.1021/mz300522n.

CHAPTER FOUR

Dendronized Systems for the Delivery of Chemotherapeutics

Lance Dockery, Marie-Christine Daniel[1]
Department of Chemistry and Biochemistry, University of Maryland Baltimore County, Baltimore, MD, United States
[1]Corresponding author: e-mail address: mdaniel@umbc.edu

Contents

Abstract

This chapter reviews the use of dendronized systems as nanocarriers for the delivery of chemotherapeutic drugs. Dendronized systems include dendrimers prepared through convergent methods as well as other systems containing dendrons (e.g., polymers, nanoparticles, liposomes). The preparation of such systems is detailed, followed by the various conjugation techniques used for the transport of chemotherapeutic drugs and their specific delivery to tumor cells. In addition, the ability of dendronized systems to provide passive and active targeting to tumors is discussed. The efficacy of drug

Advances in Cancer Research, Volume 139
ISSN 0065-230X
https://doi.org/10.1016/bs.acr.2018.04.003

delivery using dendronized systems is also illustrated through specific examples of kinetic and biological studies. Finally, the newest trends in conjugation of the most common chemotherapeutics to dendronized systems are described. Overall, this chapter highlights dendronized systems as a way to improve the therapeutic efficiency of drugs for the treatment of cancer. All the recent developments in areas, such as biodegradable dendrimers, modifications to enhance biocompatibility, selectively cleavable drug conjugations, ligand-mediated targeting, and the potential for multifunctional properties, show promises for future advances in cancer therapy.

1. INTRODUCTION

1.1 Dendronized Systems

1.1.1 General Description of Dendrimers and Dendronized Systems

Dendrimers are a class of symmetrical and highly branched molecules with a central core that radiates outward, growing at various branching points to give the overall molecule which presents the features of a tree (Gillies & Frechet, 2005) (the Greek word for tree is *dendron*). Dendrimers have become a popular area of research due to their promising biological applications in areas such as drug and gene delivery, and as molecular imaging agents (Boas & Heegaard, 2004; Lee, MacKay, Frechet, & Szoka, 2005; Menjoge, Kannan, & Tomalia, 2010). The term "generation" is commonly used to describe dendrimers based on the number of branch points encountered as you move outward from the central core (Fig. 1A). The first generation is termed generation 1 (G-1), the second generation is G-2,

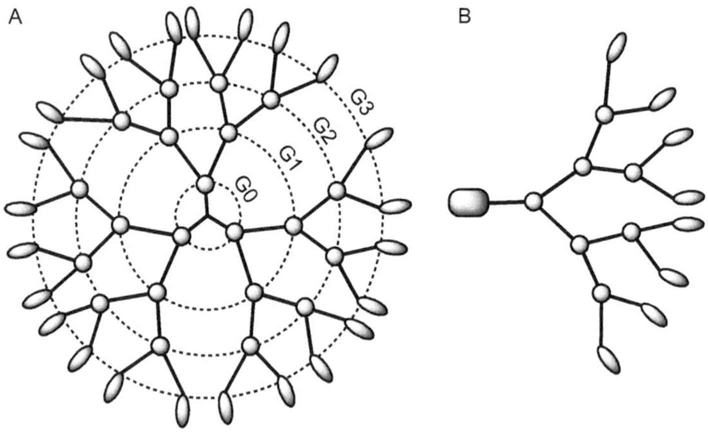

Fig. 1 (A) Schematic representation of a G3-dendrimer with branch-like architecture and (B) representation of a G-3 dendron.

and so on until the terminal branches are reached. Classical dendrimers are spherically symmetrical structures with a polyfunctional organic molecule as a core and identical end groups. Dendronized system is a more general term that includes classical dendrimers but also any system that use different types of cores and dendrons (asymmetrical conically shaped with a discrete core focal point, Fig. 1B) connected to this core.

While the generation is directly related to the number of branching points, it says nothing about the chemical nature of a dendronized system, the individual subunits making up each branched generation. Many types of dendronized systems exist (Medina & El-Sayed, 2009). The most common types of dendronized systems are derived from dendrimers depicted in Fig. 2: poly(amidoamine) (PAMAM, **1**), poly(propyleneimine) (PPI, **2**), peptide bonds (**3**), polyether, and polyester; however, numerous other structures have been synthesized. A survey of literature shows that dendrimers and dendron-based drug delivery systems are a common application of this type of nanoarchitecture. A more detailed examination of these systems in drug delivery will follow in the next section. Some other dendronized polymers also exist that help to introduce the properties of dendronized systems.

1.1.2 Other Dendronized Systems

Block dendritic (Wurm & Frey, 2011) and dendrigraft (Teertstra & Gauthier, 2004) copolymers represent two unique types of dendritic systems which show promise in the field of drug delivery. Dendritic block copolymer systems are usually in the form of either a linear polymer attached to a symmetrical dendritic structure (linear-dendritic copolymer) or a linear polymer attached to a hyperbranched unsymmetrical polymer unit (linear hyperbranched copolymer) (Fig. 3A and B).

Two applications of block dendritic and dendrigraft copolymers follow. These examples introduce one of the key advantages dendritic systems offer

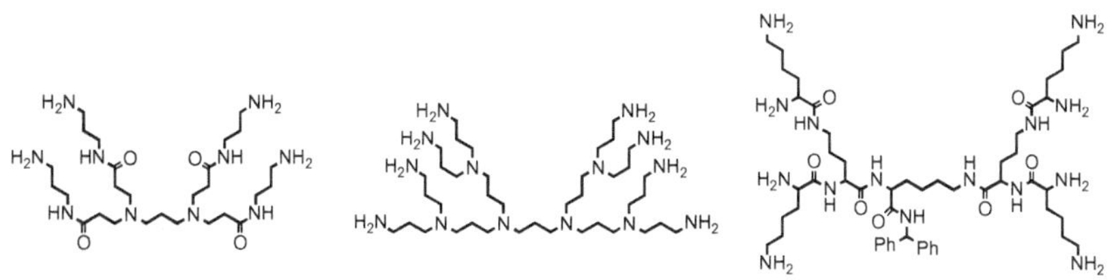

1 2 3

Fig. 2 Selection of common dendrimers: poly(amidoamine) (PAMAM) **1**, poly (propyleneimine) (PPI) **2**, and poly-L-lysine **3**.

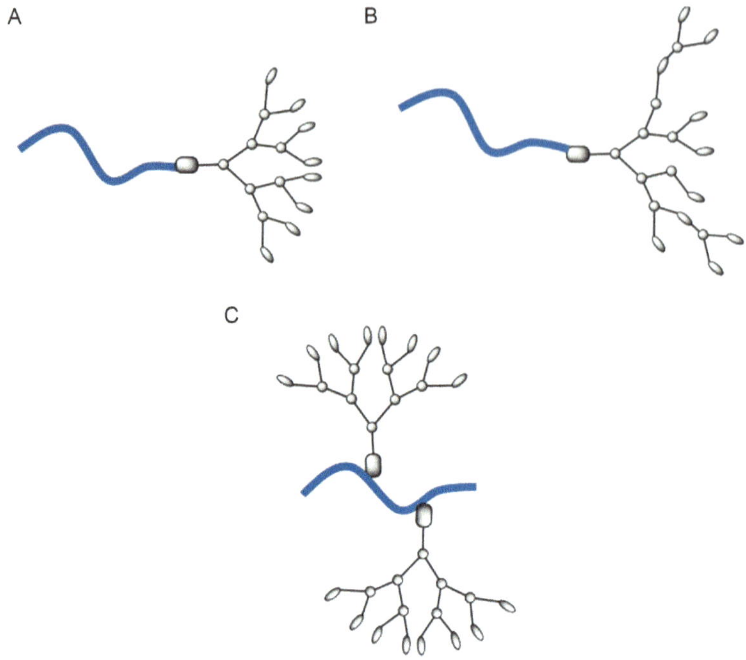

Fig. 3 Common dendritic polymer systems. (A) linear-dendritic copolymer, (B) linear hyperbranched copolymer, and (C) dendrigraft copolymer.

in the field of drug delivery: providing a vehicle to efficiently administer hydrophobic or otherwise cytotoxic therapeutics in a controlled and localized manner, contrasting traditional nonspecific delivery observed with most cancer treatments (Mintzer & Grinstaff, 2011). D'Emanuele et al. (Zhou, D'Emanuele, & Attwood, 2013) reported the synthesis of a linear-dendritic block copolymer derived from a G-2 PAMAM dendrimer and poly(butylene oxide)–poly(ethylene oxide) block copolymer. This dendritic block copolymer formed micelles which solubilized paclitaxel 3700 times more efficiently at 25°C than in a buffered solution. The dendritic block copolymer solubilized paclitaxel 8–10 times more efficiently (6 mg drug/g copolymer) than the copolymer alone. The authors went on to examine the release of paclitaxel from these micellar systems via dialysis at 37°C and found release to occur more slowly in dendritic block copolymer systems vs the copolymer only system (40%–60% release in 5 h dialysis, respectively).

Dendrigraft copolymers are traditionally synthesized by sequential grafting of dendrimers onto existing polymer chains (Fig. 3C) to give the final copolymer architecture. Gauthier et al. (Njikang, Gauthier, & Li, 2008) synthesized dendrigraft copolymer micelles and studied the release kinetics of indomethacin and lidocaine under dialysis using UV spectroscopy. In this example, poly(2-vinylpyridine) was grafted onto a polystyrene core which encapsulated the target drugs following micelle formation. The authors observed indomethacin loadings of 46%, 42%, and 43% (by weight) for G1, G2, and G3 dendrigraft copolymer systems, respectively. A low loading (only 8% in G1 by weight) was observed for lidocaine. The resulting diffusion studies gave consistent release profiles for both indomethacin and lidocaine. Free diffusion of the free drugs via dialysis (up to 100%) occurred within 5 h, while dendrigraft micelles showed a rapid initial release (70%) up to 5 h followed by a more gradual release over the following 1–2 days.

1.1.3 Factors Influencing the Polarity and Solubility of Dendronized Systems

The polarity and solubility of dendronized systems are important to their influence over the host–guest interactions of dendritic systems with potential drugs/therapeutics and on the cytotoxicity of the dendritic systems in vitro and in vivo. A more detailed discussion of how the noncovalent interactions of dendritic systems allow for drug delivery will be discussed in Section 3.2. The focus of this section will be (1) broadly, how the polarity and solubility of dendronized systems can impact cytotoxicity and (2) what common modifications are made to surface groups of dendronized systems and how those modifications can afford control over polarity, solubility, and cytotoxicity.

Under the physiological conditions, dendritic systems can display cationic, anionic, or neutral surface charge. Under physiological pH, primary amine groups become protonated and endow a cationic character. Anionic character can be conferred from carboxylates, while neutral surface groups such as ether or alcohol groups remain uncharged. Cationic dendronized systems are associated with a high degree of cytotoxicity due to their ability to interact with the negative charge of plasma membranes (Jevprasesphant et al., 2003). Conversely, anionic systems often exhibit low cytotoxicity.

Conjugation of lauroyl chains (anionic character) and polyethylene glycol (PEG) to the surface of G2–G4 PAMAM-NH_2 dendrimers reduces cytotoxicity sevenfold in Caco-2 cells compared to unmodified PAMAM dendrimers (Jevprasesphant et al., 2003). Conjugation of G-5 PAMAM surface groups to acetic anhydride (Majoros, Myc, Thomas, Mehta, & Baker, 2006; Majoros, Thomas, Mehta, & Baker, 2005) also reduced cytotoxicity in vitro and enhanced solubility in DMSO. PEGylation and carboxylation of poly(ether imine) dendrimers (Krishna, Jain, Tatu, & Jayaraman, 2005) reduced cytotoxicity on African green monkey CV-1 cells by as much as 98% at a dendrimer concentration of up to 100 mg/mL. While there is inherent cytotoxicity associated with most dendronized systems, a wide range of surface modifications have been established to overcome these limitations and make drug delivery possible.

1.2 Common Types of Dendrimers and Dendrons for Drug Delivery

1.2.1 PAMAM Dendronized Systems

Dendrimers synthesized from poly(amidoamine) subunits were the first type of dendronized system reported by Tomalia et al. (1985). Subsequent publications elaborated on this unique type of structure and provided a better understanding of their applications in drug delivery (Esfand & Tomalia, 2001; Tomalia, Naylor, & Goddard, 1990). The core of PAMAM dendrimers is usually a molecule such as ethylene diamine (G0). Successive Michael reactions with methyl acrylate can give additional generations as observed in Fig. 4. The primary amines in the outer generation can be further elaborated to carboxylates (Tomalia et al., 1985) or conjugated to other molecules to give the desired therapeutic properties (Menjoge et al., 2010) (drug delivery, imaging, active targeting, etc.).

Several properties make PAMAM dendrimers useful for drug delivery: their commercial availability (Mintzer & Grinstaff, 2011), their well-characterized synthetic reactions (Esfand & Tomalia, 2001) (including conjugation and surface group modification), and excellent water solubility (Patri, Majoros, & Baker, 2002). Dendritech Inc. offers simple G1–G10 dendrimers for purchase in addition to PAMAM dendrimers with a variety of surface functional groups. The terminal groups of PAMAM dendrimers have been functionalized with therapeutic agents through many reactions, including *N*-hydroxysuccinimide (NHS) coupling, reductive amination and "Click" reactions, among others. Examples of specific reactions will be discussed later. Some drugs which have been effectively conjugated and delivered via PAMAM dendrimers include doxorubicin (DOX) (Li et al., 2012; Zhang et al., 2011) and paclitaxel (PTX) (Cline et al., 2013; Teow et al., 2013). A more detailed discussion of the delivery of specific types of drugs is covered toward the end

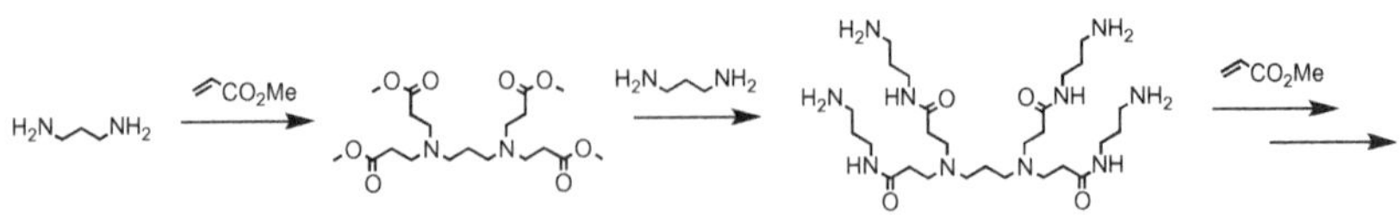

Fig. 4 Synthetic scheme for the preparation of PAMAM dendrimers.

of this chapter. A similar class of molecules are PPI dendrimers. These dendrimers are synthesized by sequential Michael additions of acrylonitrile to terminal amine groups (discussed in Section 2.1.1).

While there are many advantages to PAMAM dendrimers, some disadvantages are worth noting. As discussed earlier, the cationic nature of amine groups exhibits high cytotoxicity (Jevprasesphant et al., 2003) and surface modifications are required. Additionally, incomplete reactions can result in problems with the polydispersity of the final product and larger generation dendrimers are often challenging to work with due to steric crowding (Mintzer & Grinstaff, 2011).

1.2.2 Polyether and Polyester Dendronized Systems

Another common class of dendronized systems include those derived from ether and ester linkages (Fig. 5). Dendrimers of this type are considered to be more biodegradable (ester, **4**) or generally biocompatible (ether, **5**) and more easily tolerated within the body (Leiro, Garcia, Tomas, & Pego, 2015). They have been synthesized by a variety of divergent (Ihre, De Jesus, & Frechet, 2001) or convergent (Gillies & Frechet, 2002) methods. A more monodisperse dendrimer is often obtained by these methods comparable to PAMAM and similar dendrimers. Dendrimers containing polyester linkages have been studied for the delivery of doxorubicin (De Jesus, Ihre, Gagne, Frechet, & Szoka, 2002; Guillaudeu et al., 2008; Ihre, De Jesus, Szoka, & Frechet, 2002; Lee et al., 2006) both in vitro and in vivo and the effective encapsulation of camptothecins (Morgan et al., 2006).

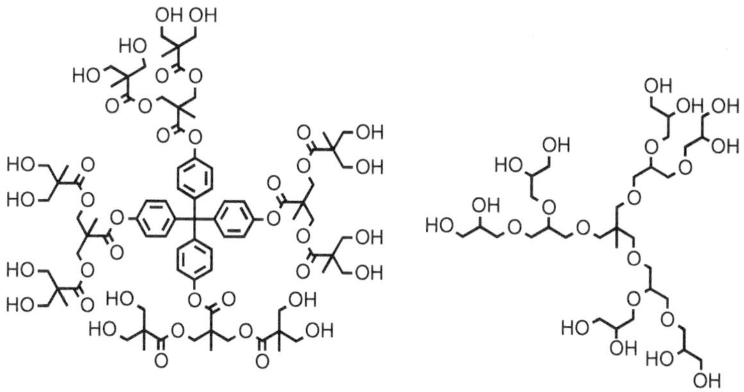

4 **5**

Fig. 5 Representations of common polyester (**4**) and polyether (**5**) dendritic systems.

Like PAMAM dendrimers, polyester and polyether dendrimers are often difficult to synthesize in higher generations. Other challenges associated with this class of molecules include poor water solubility and premature cleavage by nonspecific enzymes of conjugated therapeutic agents before they arrive at the target tissue.

2. SYNTHESIS OF DENDRONIZED SYSTEMS

2.1 Conventional Dendrimer Synthesis

2.1.1 General Divergent Synthesis

The earliest dendrimers, including those by Tomalia et al. (1985, 1990) and Newkome, Yao, Baker, and Gupta (1985), were synthesized via a divergent route. This refers to a process through which a single multifunctional core reacts with sequential rounds of monomer units, giving rise to dendrimers of increasing generations. In each generation, after the addition of one monomer unit the structure diverges (Fig. 6) with the terminal functional group able to react again. PAMAM and PPI dendrimers are the most common dendrimers synthesized by divergent means. In the case of PAMAM (Fig. 4), ethylene diamine (EDA) is reacted with methyl acrylate via Michael addition to give a carbonyl end group which is reacted again with the EDA and the sequence repeats. For PPI dendrimers, EDA or a similar molecule is reacted with acrylonitrile. The nitrile is reduced, and the sequence is repeated with additional acrylonitrile (Pan, Grow, Wilson, & Daniel, 2013) to reach the desired generation of dendrimer.

In general, these divergent reactions suffer from several drawbacks, the biggest of which being the large excess of methyl acrylate or similar molecules that must be used to drive the reaction to completion (Mintzer & Grinstaff, 2011). A purification step must usually be done after each generation is synthesized. This process can be tedious and/or difficult in some cases depending

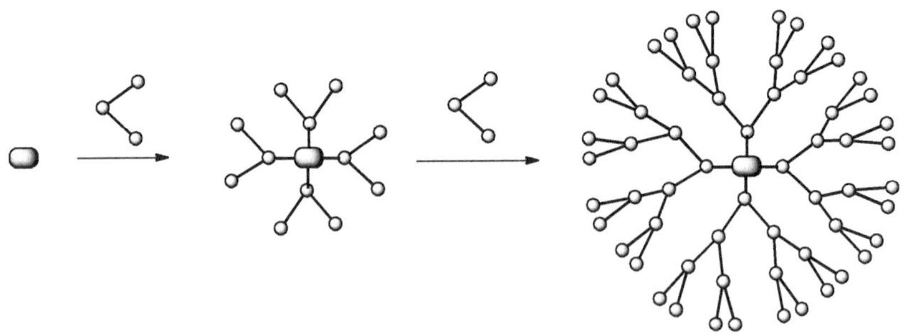

Fig. 6 Scheme of a divergent synthesis of a dendritic system.

on the dendrimer functionality. In addition, many side reactions can give rise to imperfect dendrimers. Even with a large excess of the Michael addition monomer, incomplete reactions can occur (Carlmark, Hawker, Hult, & Malkoch, 2009). In other cases, retro-Michael reactions and intramolecular cyclizations may take place. All these side reactions increase in likelihood as the dendrimer generation increases, making monodispersity a significant challenge for higher generation molecules.

Despite these problems, dendronized systems prepared via divergent synthesis routes (conventional dendrimers) are commonly used for drug delivery. The ability to synthesize these dendrimers and control cytotoxicity through functional group conversion as previously noted is a significant advantage, which is reflected by the vast number of publications detailing divergent methods (Ihre et al., 2001; Newkome & Shreiner, 2008). Often, however, the divergent method alone is insufficient for synthesizing complex dendronized systems. A detailed examination of specific methods for preparing dendronized drug delivery systems will be discussed in Section 3.

2.1.2 "Click" Chemistry in Conventional Dendrimers

With the development of "Click" chemistry by Sharpless et al. (Kolb, Finn, & Sharpless, 2001), dendronized systems can be prepared in a more efficient manner, requiring minimal purification (Wu et al., 2004). Hawker et al. demonstrated this approach in 2005 to synthesize third-generation azide dendrimers using copper catalysts (Joralemon et al., 2005; Wu et al., 2005). In these examples, the "Click" reaction was used at the branch points and the attachment of two fragments at the focal point. The same group reported that the thiol-ene "Click" reaction could also be used to synthesize up to G4 poly(thioether) dendrimers (Killops, Campos, & Hawker, 2008). While "Click" chemistry is a powerful tool for the synthesis of divergent systems, it has the greatest potential in the synthesis of complex dendronized systems, those synthesized convergently with multifunctional properties (vide infra) (Franc & Kakkar, 2010).

2.2 Synthesis of Complex and Multifunctional Dendronized Systems

2.2.1 Convergent Synthesis of Dendrimers

The convergent growth method involves synthesizing separate dendrons (usually via divergent methods) and then coupling them together or to a common core to form the overall dendrimer (Fig. 7) (Medina & El-Sayed, 2009). When coupling them together, the dendrons "focal points are not activated

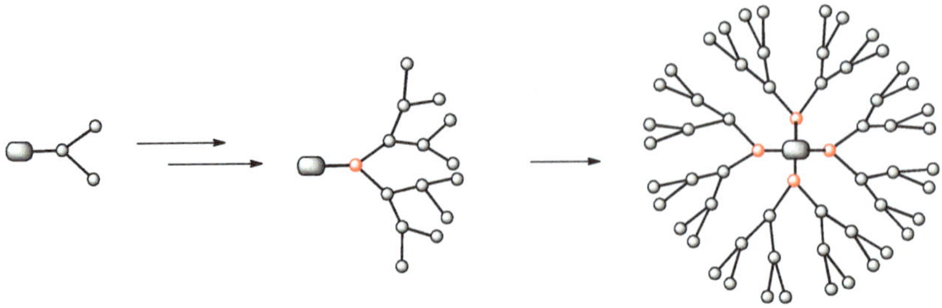

Fig. 7 Scheme of a convergent synthesis of a dendritic system.

until specific conditions are met to bring the separate dendrons together into one molecule (e.g., "Click" chemistry). The challenge with this method is that the sterics of the dendrons can severely limit the effectiveness of the coupling reaction. When attaching them to a common core, the dendrons" focal point needs to be reactive with a least part of the core. The final molecule must be purified but this is often easier (i.e., dialysis) than other methods due to the significant difference in size of the dendrons and the final dendronized system.

The main advantage to convergent synthesis is that it allows for differently functionalized dendrons to be coupled, leading to complex and multifunctional architecture. Dendrimers can be designed with drug delivery, magnetic resonance imaging or other contrast agents, photodynamic and photothermal therapy, and ligand-based targeting in mind to enhance the overall therapeutic utility of dendronized drug delivery systems (Mintzer & Grinstaff, 2011). This section will be focused primarily on the synthesis of multifunctional drug delivery systems.

2.2.1.1 Synthesis of Janus Dendronized Systems

"Janus" dendrimers are a class of molecules featuring two unique dendritic units joined together at a single focal point. In this way, a complex architecture can be assembled through the approach depicted in Fig. 8. Fragments are synthesized independently and brought together either through a single reaction at the focal points or by attachment to a polyfunctional core molecule. This method was demonstrated in several early examples by Frechet et al. (Hawker & Frechet, 1990; Wooley, Hawker, & Frechet, 1991), where three symmetrical polyester dendrons were joined together around a trifunctional core initiator. Frechet et al. then took this method a step further by developing the "Bow-tie" dendrimers (Gillies & Frechet, 2002; Lee et al., 2006). In the "bow-tie" dendrimers, two distinct dendritic structures, which were synthesized through a hybrid convergent/divergent method,

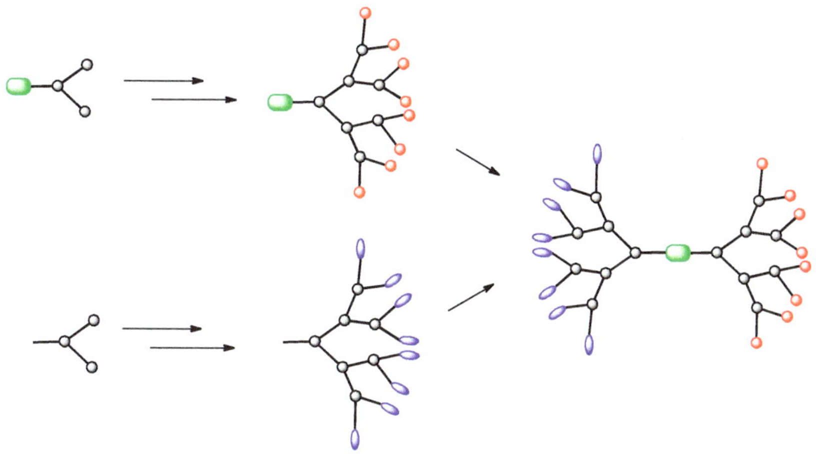

Fig. 8 Schematic representation of a convergent–divergent approach to synthesize multifunctional Janus type or dendrimers.

were connected at a common focal point (Fig. 8). This method afforded dendrimers with multifunctional potential, a property which would be the focus of much of the work to follow.

Sun et al. synthesized Janus-type dendrimers which could assemble into micelles for the delivery of doxorubicin and subsequent drug release induced and monitored with NIR irradiation (Sun, Ma, Dong, Zhu, & Zhu, 2012). Acton et al. reported the synthesis of a class of PEG-based Janus dendrimers bearing two separate model drugs in a convergent manner, allowing precise control over the loading ratio of the two drugs to be delivered (Acton et al., 2013). Acton demonstrated a powerful application of "Click" chemistry in the preparation of dendronized systems: two separate amphiphilic or multifunctional dendritic parts were easily joined via a [3+2] cycloaddition reaction to give a triazole core from corresponding alkyne and azide (Arseneault, Wafer, & Morin, 2015).

2.2.1.2 Synthesis of Multifunctional Dendronized Systems

With the advance of more complex bow-tie and Janus dendrimers, an effort was made to include a wide array of therapeutic properties in these systems. Baker et al. pioneered this field, with the initial G5-PAMAM dendrimer system incorporating acetylated primary amino groups (solubility) and conjugated fluorescein isothiocyanate (imaging), folic acid (FA, targeting), and methotrexate (MTX, drug) into one molecule which induced cytotoxicity to KB cells in vitro (Majoros et al., 2006, 2005; Thomas et al., 2005).

Internalization of the dendrimer could be observed via confocal microscopy and cellular uptake was promoted by FA but while the dendrimer showed antiproliferative properties in KB cells, the activity was reduced compared to free MTX (Thomas et al., 2005). Despite the novel design of these multifunctional dendrimers, the work by Baker et al. highlights the need for (i) better control over the conjugation chemistry (importance of selectively cleavable bonds), (ii) optimized loading of more drugs per dendrimer, and (iii) larger dendronized systems with higher drug loading capacity as delivery vehicles.

Additional advances in the use of "Click" chemistry for the synthesis of complex dendrimers were pioneered by Marcus Weck at NYU. Initially only monofunctionalization was possible on PAMAM dendrimers (Yoon, Goyal, & Weck, 2007). However, the group later improved upon the cycloaddition protocol to eliminate the use of toxic copper catalyst using a strain-promoted reaction in which complete functionalization of the dendrimer termini was possible (Ornelas, Broichhagen, & Weck, 2010). A more complex version of this architecture was then reported containing three unique termini: (Ornelas, Pennell, Liebes, & Weck, 2011) the final dendrimer contained 9 azide termini, 9 amine termini, and 54 terminal carboxylate groups. To prove the practical utility of this dendronized system, conjugation to a NIR fluorescent dye was reported and cytotoxicity tested against T98G cells. No significant cytotoxicity was observed even after 72 h (up to 50 μM). The emerging trend in designing dendronized drug delivery systems is a need for not only efficient drug loading and delivery but also multifunctional systems which can accommodate other functions such as imaging and active targeting.

2.2.2 Synthesis of Hybrid Dendronized Systems

Combining the drug-delivery potential of dendronized systems with gold nanoparticles to form hybrid systems has been an area of emerging research over recent years (Brunetti, Bouchet, & Strumia, 2015). Daniel et al. reported the synthesis of G3-PPI dendrons with nitrile and amine terminal groups (Daniel et al., 2011). These dendrons were coordinated onto 18-nm gold nanoparticle cores through a tetraethylene glycol (TEG) spacer with a thiol end group. The resulting dendronized nanoparticles were characterized by UV–vis and FT-IR spectroscopy, dynamic-light scattering (DLS), zeta potential, and transmission emission microscopy (TEM). The authors also evaluated the stability of these dendron-capped nanoparticles and found them to be stable in a variety of solution conditions (buffers, high salt concentrations, various pHs) as well as to withstand freeze-drying.

Cho et al. synthesized four Dendron-Au nanoparticles containing G1-CO_2H, G2-CO_2H, HS-G1-CO_2H, and TA-CO_2H PAMAM dendrons (Cho, Zangmeister, MacCuspie, Patri, & Hackley, 2011). The gold colloids used to react with the dendrons were usually 10 nm diameter, and the resulting nanoparticles were characterized by NMR, MALDI-TOF, DLS, UV–vis, and X-ray photoelectron spectroscopy. The stability of these nanoparticles was also examined over a range of temperatures and pH values in addition to chemical stability toward KCN. Several other similar stability studies have been reported for Au-dendrimer systems (Deol, Weerasuriya, & Shon, 2015; Tsai, Cho, Elzey, Gigault, & Hackley, 2013), and Kim et al. (2012) described the application of dendronized gold nanoparticles for the delivery of siRNA therapeutics.

Li et al. reported the synthesis of a dendrimer-based gold nanoparticle as a bimodal imaging agent for magnetic resonance (MRI)/computed tomography (CT) imaging (Chen et al., 2013; Wen et al., 2013) of breast cancer in vitro and in vivo (Li et al., 2013). The authors functionalized commercial G5-PAMAM dendrimers with a gadolinium contrast agent (DOTA-NHS) and polyethylene glycol monomethyl ether chains (solubility). They then entrapped 4-nm gold nanoparticles inside the dendritic cavities. The resulting dendrons and nanoparticles were characterized by NMR, UV–vis spectroscopy, TEM, and DLS. The resulting Gd—Au nanoparticles-containing dendrimers were noncytotoxic up to a concentration of 50 μM after 24 h in vitro. Mouse xenografts with MCF-7 tumors were examined over 24 h following injection of the Gd—Au nanoparticles with the highest signal-to-noise ratio observed at 120 and 100 min, respectively, for MRI and CT imaging (Li et al., 2013).

3. DENDRONIZED SYSTEMS FOR DRUG DELIVERY

3.1 Covalent Conjugation

In dendronized systems, covalent conjugation and noncovalent interactions represent the two primary methods to incorporate drugs for delivery. Both methods will be discussed, however, the focus of this section will be an overview of covalent conjugation methods. Examples of the most common covalent conjugation methods will be given, however, other techniques do exist (Kaminskas, McLeod, Porter, & Boyd, 2012; Wu, Ficker, Christensen, Trohopoulos, & Moghimi, 2015). The enhanced permeation and retention (EPR) effect will also be briefly discussed as it pertains to the uptake and distribution properties of drugs conjugated on dendronized systems.

3.1.1 Hydrazone Bond Conjugation

The acyl hydrazone bond is an acid-labile bond resulting from the conjugation of a hydrazine moiety to a carbonyl group (Binauld & Stenzel, 2013). This bond, among other acid-labile types, can be used to deliver drugs based both on the relative acidic environment of tumors (6–7) (Gerweck & Seetharaman, 1996) compared to normal tissues and the low lysosomal pH within the cell (pH ~ 4–5). The acyl hydrazone bond is featured prominently in drug delivery systems where the drug is conjugated to the termini of a dendronized system. One of the earliest examples comes from De Jesus et al., where a 3-arm poly(ethylene oxide) dendrimer was designed to incorporate doxorubicin via a hydrazone bond (De Jesus et al., 2002). This dendritic system showed ideal release of doxorubicin as pH decreased over time (Fig. 9). This use of the hydrazone bond was reported in several other early publications and has been reported frequently since for the delivery of several different therapeutic molecules such as methotrexate and adriamycin (Dhanikula, Argaw, Bouchard, & Hildgen, 2008; Ihre et al., 2002; Kono et al., 2008).

3.1.2 Ester Bond Conjugation

Drug conjugation through ester bonds is a less common but still a feasible method as lysosomal esterase activity will favor hydrolysis of drug conjugates and release of the drug molecule. D'Emanuele et al. evaluated the stability of amide- and ester-conjugated naproxen prodrugs to G0-PAMAM

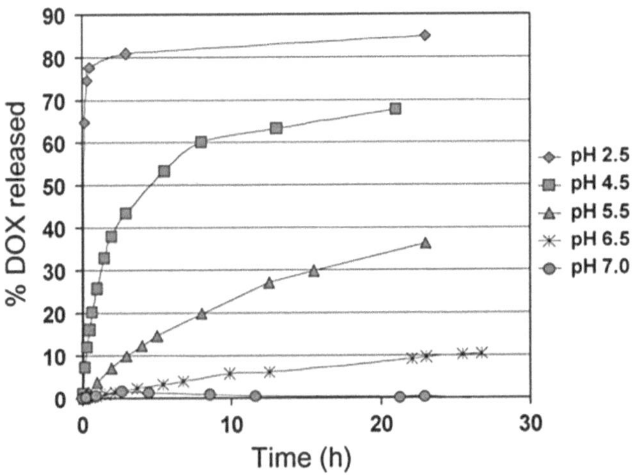

Fig. 9 pH profile for release of doxorubicin from conjugate. *Reprinted with permission from De Jesus, O. L. P., Ihre, H. R., Gagne, L., Frechet, J. M. J., & Szoka, F. C. (2002). Polyester dendritic systems for drug delivery applications: In vitro and in vivo evaluation. Bioconjugate Chemistry, 13(3), 453–461. Copyright 2002 American Chemical Society.*

dendrimers under a variety of pH conditions (Najlah, Freeman, Attwood, & D'Emanuele, 2006). Amide and ester conjugates were stable over a pH profile of 1.2–8.5; however, the esters were selectively cleaved enzymatically. As discussed previously, Baker et al. conjugated MTX to G5-PAMAM dendrimers (five molecules per dendrimer) via cleavable ester bonds along with targeting (FA) and imaging (FI) functionality. The multifunctional ester-conjugated MTX dendrimer was found to 100-fold more active than free MTX in vitro (Majoros et al., 2005; Quintana et al., 2002).

3.1.3 Peptide Bond Conjugation

Enzymatic cleavage of specific peptide sequences represents another potential route for specific drug delivery. Kurtoglu et al. synthesized PEGylated-G4-PAMAM-ibuprofen conjugates via a GFLG-peptide linker with between four and seven molecules per dendrimer as characterized by HPLC (Kurtoglu, Mishra, Kannan, & Kannan, 2010). Drug release was achieved in a solution of cathepsin B in 80% (v/v) human plasma, with 65% of payload released within 10 h compared to a simple amide conjugate which showed only 40% payload release at 48 h and an ester conjugate which had no drug release under these enzymatic conditions. Additional examples using similar GFLG peptide systems will be discussed later (Li et al., 2014; Zhang et al., 2014).

3.1.4 "Click" Conjugation

"Click" reactions are useful for synthesizing multifunctional dendronized systems but also for conjugating drugs to the end groups of dendrimers, particularly due to the high atom economy of this reaction and minimal side reactions. While a copper catalyst is often required under standard conditions, copper-free "Click" reactions are also possible, thus eliminating the need to remove toxic copper from the products via purification. Feng et al. reported the synthesis of G1–G3 Janus-type poly(ethylene oxide) dendrimers functionalized with FA and camptothecin (CPT) via copper-catalyzed "Click" chemistry (Feng, Pinaud, Chaikof, Taton, & Gnanou, 2011). The authors observed full conjugation of FA and CPT to PEO dendrimers and characterized the products via ^{1}H NMR spectroscopy. Huang et al. reported the conjugation of MTX to PAMAM dendrimers without the need of a copper catalyst (Huang et al., 2011). G5-PAMAM dendrimers were first functionalized with cyclooctyne and unreacted primary amines protected via acetylation. The cyclooctyne–dendrimers were then reacted with azide-functionalized MTX under mild conditions (methanol, 24 h, rt). The products were characterized by ^{1}H NMR spectroscopy and HPLC and the data indicated 5, 10, and 17 conjugated MTX molecules per dendrimer.

3.2 Noncovalent Conjugation

Noncovalent interactions between the drug and a dendronized system (often referred to as host–guest interactions) may arise from ionic interactions, hydrogen-bonding interactions, and encapsulation (hydrophobic interactions). The structure of the drug to be noncovalently conjugated plays a significant role in loading efficiency as do factors such as dendrimer size, internal and surface modifications, and solution pH and solutes.

PAMAM dendrimers are a common model system to study their host–guest interactions with a variety of drugs. Under conditions where primary amines are protonated, ionic interactions (with an acidic guest molecule) predominate. Systems with alcohol surface functionalization will have hydrogen-bonding interactions, and less polar dendrimers (such as those functionalized with alkyl esters) will interact primarily through hydrophobic encapsulation. Chauhan et al. demonstrated this trend by examining the encapsulation potential of indomethacin in G4-PAMAM dendrimers at various pH levels (Chauhan, Jain, Diwan, & Khopade, 2004). The aqueous solubility of camptothecin can be improved up to 20-fold with G4–G6 PAMAM dendrimers alone via encapsulation (Cheng, Li, & Xu, 2008). In another study, encapsulation of mycophenolic acid (MPA) by G3–G5 PAMAM dendrimers was characterized by detailed ^{1}H and NOESY NMR experiments (Hu, Cheng, Ma, Wu, & Xu, 2009). The structural studies showed evidence for encapsulation occurring both through hydrophobic encapsulation and ionic interactions with charged surface groups and MPA.

Other factors such as salt concentration and the presence of albumin or other serum proteins can also influence the extent of host–guest interactions (Prajapati, Tekade, Gupta, Gajbhiye, & Jain, 2009). For molecules utilizing noncovalent interaction, individual testing is needed to determine the effectiveness of covalent vs noncovalent conjugation. In one example, Patri et al. studied the release kinetics of MTX for a G5-PAMAM dendrimer in covalent and noncovalent conjugation conditions (Patri, Kukowska-Latallo, & Baker, 2005). They found that covalent delivery of MTX was enhanced over free drug, especially for systems with active targeting and or multifunctional properties.

3.3 EPR Effect, Tumor Microenvironment, and Drug Delivery

While optimal drug delivery is often achieved by dendronized systems possessing active targeting methods, it is useful to point out that the EPR effect and the microenvironment around a tumor also allow for passive targeting.

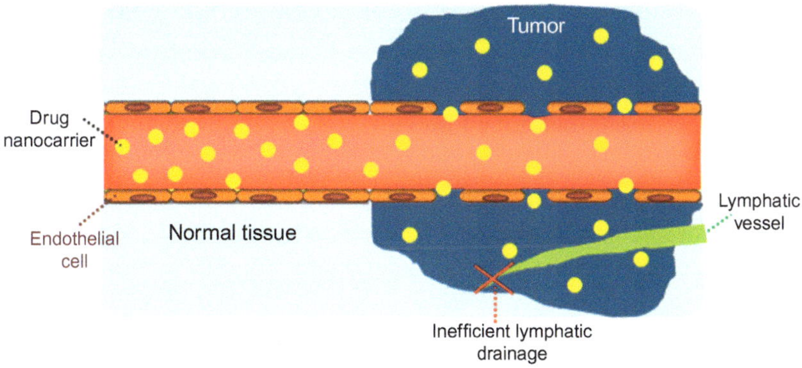

Fig. 10 Enhanced permeability and retention (EPR) effect: Passive targeting. *From Daniel, M.-C. (2012). Drug delivery carriers. In J. R. Lakowicz & J. Zhang (Eds.), Emerging applications of colloidal noble metals in cancer nanomedicine (pp. 54–67). Future Medicine, London, UK.*

The EPR effect was first discovered by Matsumura and Maeda (1986). Due to the rapid proliferation of a tumor, the surrounding vasculature can be seen as "leaky," creating susceptibility to penetration by dendronized systems and other nanoparticles of sufficiently small size (Fig. 10) (Daniel, 2012). In normal, healthy vasculature, the dendronized systems, or nanoparticles are too large to penetrate the epithelial cell walls. Thus, accumulation will be expected in higher levels in tumor tissues over healthy ones, assuming sufficient blood circulation and pressure.

4. KINETIC AND BIOLOGICAL STUDIES

4.1 Drug Release

Understanding how temperature, pH, and other external factors influence drug release allows for optimizations of the kinetics of drug release. Applying these concepts to in vitro and in vivo studies in turn allows design of dendronized drug delivery systems for biological settings (Menjoge et al., 2010). As mentioned in previous studies, the low pH of the tumor microenvironment and lysosome can trigger release of drug from acid-labile bonds such as the acyl hydrazone. For evaluation of intravenous injection, examining drug release in PBS buffered (pH 7.4) solutions may be used. For example, Singh et al. designed a Folate-PEG-PAMAM dendrimer for the delivery of 5-fluorouracil (5-FU) and tested the drug release both in vitro and in vivo (Singh, Gupta, Asthana, & Jain, 2008). Studies of the in vitro release via dialysis of the 5-FU loaded dendrimer (up to 31.5% w/w) in water showed 40% drug release (UV–vis measurement) at 8 h followed by a plateau.

In another study mentioned previously, amphiphilic dendrimers were synthesized (via "Click" chemistry) and assembled into doxorubicin-containing micelles (Sun et al., 2012). Upon light irradiation, the nonpolar diazonaphthoquinone underwent a Wolff rearrangement, reversing the polarity of the end of the dendrimer and inducing the release of doxorubicin. Using this model, the in vitro release profile of this drug delivery system could be studied in more detail. Based on UV–vis measurements, the DOX loading in assembled micelles was 8.3–15.6 wt%. These DOX-nanoparticles were placed in dialysis tubing (PBS buffer, pH 7.4) and irradiated (808 nm) for 0, 10, and 30 min. The doxorubicin release was measured via UV–vis concentration over the course of 400 h. Without irradiation, full release (~95% plateau in all cases) occurred after 400 h. Ten-minute irradiation resulted in full release after 200 h and 30 min irradiation, full release after 100 h. In this case release of the drug can be controlled to an extent simply by irradiation duration and the reversed polarity of the dendrimer.

4.2 Drug Distribution

4.2.1 EPR Effect/Passive Targeting

Achieving proper drug distribution in a passive manner requires taking advantage of the EPR effect (Matsumura & Maeda, 1986). As mentioned earlier, the EPR effect takes advantage of the leaky epithelial walls of blood vessels surrounding tumor sites and subsequent increased permeation by nanoparticle systems of appropriate size. However, there are several factors which can affect the efficiency of passive targeting in biomedical applications (Nichols & Bae, 2014). Tumors are highly heterogeneous and uniform distribution does not occur via EPR-mediated passive targeting (Maeda, 2015). Variable size and vascularization can also hinder efficient penetration of drug delivery systems into tumor tissue. Nonetheless, PEGylated dendronized systems often do exhibit longer blood circulation times compared to traditional anionic dendrimers and good biocompatibility as was observed by Kaminskas et al. when ^{3}H-labeled poly-L-lysine dendrimers were administered to rats intravenously (Kaminskas et al., 2008). Overall, delivery of dendronized systems via passive targeting is insufficient alone to enable effective cancer treatment and multifunctional delivery systems which can benefit from EPR but also display active targeting moieties are ideal.

4.2.2 Ligand-Based Active Targeting

Ligand-based active targeting takes the form of specific ligands or antibodies recognized by overexpressed receptors or antigens found on the surface of tumor cells. In this manner, the therapeutic efficacy of nanoparticle-based drug delivery systems is greatly improved. The most commonly used ligands include folic acid (FA), biotin, transferrin (Tf), aptamers, and monoclonal antibodies. This section will provide a brief overview of specific examples in which active targeting improved drug delivery in dendronized systems, many examples in the following section discussing recent advances in the delivery of specific drugs will also contain active targeting molecules.

Li et al. designed a G4-PAMAM-PEG dendrimer functionalized with an acid-labile doxorubicin bond (acyl hydrazone) and transferrin on the surface (Li et al., 2012). Tamoxifen (TAM) was also included in the interior of the dendrimer as this type of dendrimer has been shown to improve blood–brain barrier penetration, conferring it potential in treatment for brain gliomas. The authors found the drug loading to be 7 doxorubicin molecules, 1 transferrin, and 29 TAM molecules encapsulated per dendrimer (DOX loading characterized by UV–vis and Tf characterized by SDS-PAGE). The acid-labile bond showed 6% doxorubicin release at pH 7.4 and 32% at pH 4.5. In vitro experiments showed enhanced blood–brain barrier penetration of the G4-DOX-PEG-Tf-TAM dendrimer and a continued decrease in glioma tumor volume when monitored over 48 days.

In a similar paper by Zhang et al., G4-PAMAM-PEG dendrimers were also functionalized with doxorubicin, however, in place of transferrin, a peptide sequence of Arg-Gly-Asp (RGD) was conjugated to the dendrimer to aid in active targeting (Zhang et al., 2011; Zhu et al., 2011). This peptide sequence was observed to mediate cellular binding and uptake in other instances (McNerny et al., 2009; Waite & Roth, 2009). Each dendrimer was found to contain 19 molecules of PEG, 10 molecules of RGD, and 13 molecules of DOX (attached though *cis*-aconityl acid-labile bond), as determined by HPLC. The final G4-DOX-PEG-RGD conjugates were examined using both in vitro and in vivo methods. In vitro cytotoxicity of the acid-labile dendrimer was higher than acid-insensitive doxorubicin conjugates and RGD-modified dendrimers exhibited higher cytotoxicity than dendrimers without RGD. In vivo tests on mice with

G4-DOX-PEG-RGD dendrimers revealed higher accumulation of active-targeting dendrimers than those without RGD and prolonged survival of the tumor-bearing mice (40%–50% after 60 days vs 0% after 20 days with saline and DOX alone).

4.3 Therapeutic Efficiency

4.3.1 Cellular Assays

Evaluation of the therapeutic efficiency of dendronized drug delivery systems is essential to the development process. Cellular assays are often the easiest mode and first stage to test the cytotoxicity and antitumorigenic properties of a system, and some highlights of cellular models used for this purpose will be mentioned here.

Papagiannaros et al. incorporated G4-PAMAM-DOX dendrimers (PAMAM:DOX molar ratio of 3:1, measured by UV–vis to be 97% incorporated in the dendrimer) into several different liposomes (Papagiannaros, Dimas, Papaioannou, & Demetzos, 2005). The final G4-PAMAM-DOX-liposomes had an average size of 115 nm. Compared to liposomal formulations containing only doxorubicin the authors found the release from the dendrimer–liposome nanoparticle to be slower and more controlled. The authors examined cytotoxicity against a broad range of cancer cell lines (lung, breast, colon, and prostate cancers, and central nervous system). The cancer cell lines were treated with up to 20 μM of G4-PAMAM-DOX-liposomes and monitored for in vitro growth inhibition over the course of 48 h (growth inhibiting concentration 50; total growth inhibition; lethal concentration 50). All cancer cell lines examined experienced significant growth inhibition. The GI_{50} was as low as 1.3 μM in one cell line (lung, NCI-H460) and as high as 15.4 μM in another (CNS, SF268) for the best liposomal–dendrimer–DOX formulation.

In another experiment, Al-Jamal et al. synthesized cationic poly-L-lysine dendrimers which could complex with doxorubicin and tested these dendrimers in several in vitro and in vivo experiments. The authors sought to complex doxorubicin noncovalently through hydrogen-bonding interactions with the poly-L-lysine dendrimer (DM), and while the ratio was relatively low (molar ratio of 1:10, DOX:DM characterized by UV–vis and ^{1}H NMR), the system showed strong potential so a series of cytotoxicity assays were performed. Examination of the penetration capacity of the DM–DOX complex toward prostate cancer (DU145) tumor spheroids found enhanced penetration of the spheroids compared to free DOX.

Additional experiments such as flow cytometry (to measure apoptotic cells) and tumor growth inhibition tests confirmed the enhanced cytotoxicity of DM–DOX complexes over free DOX.

4.3.2 Mouse Models

Regarding in vivo testing of dendronized systems, many publications examine xenograft models and are primarily interested in monitoring tumor volume reduction (see Section 5). There is, however, also interest in the kinetics and drug distribution properties of dendrimers in vivo. This section will highlight two studies looking at biodistribution and kinetics of dendronized drug delivery systems in mouse models.

Jain et al. synthesized G4.5-PPI-CO_2H dendrimers (characterized by FT-IR, ^1H NMR, TEM, and flow cytometry) which noncovalently encapsulated paclitaxel (PTX) (4:1 M ratio of PTX:dendrimer) and were surface-conjugated with an antibody (3:1 M ratio of Ab:dendrimer) complementary to mesothelin, a protein overexpressed in ovarian cancer (Jain, Tare, Mishra, & Tripathi, 2015). The resulting in vivo pharmacokinetic studies found the biological half-life of these dendrimers to be prolonged fourfold over free PTX, and prolonged twofold over similar G4.5-PPI-CN dendrimers. This corroborates the general enhanced biocompatibility of anionic dendrimers discussed earlier. Next, organ distribution studies were performed on these drug conjugates. The Ab-targeted dendrimer showed higher uptake in tumor tissue ($55.25 \pm 2.14\,\mu g/g$ tissue up to 24 h) compared to Ab-free dendrimers ($36.87 \pm 1.11\,\mu g/g$) indicating increased specificity afforded by the targeting moiety.

In a work similar to some mentioned earlier, Li et al. synthesized PEGylated G4-PAMAM-DOX conjugates (9.62 wt% DOX loading) using a selectively cleavable GFLG-peptide linker (Li et al., 2014). The authors performed experiments to measure both their in vivo toxicity to various organs as well as their biodistribution in mouse models. Tumor growth was compared over 21 days to a saline control, free DOX, and the dendrimer conjugate. Tumor volume was increased by 2000% in the control, 950% in free DOX, and only 500% in the conjugated system. Compared with free DOX, the PAMAM-GFLG-DOX conjugates induced lower toxicities and showed low distribution in normal organs of tumor-bearing (4T1, murine breast cancer) mice compared to healthy ones. Heart, liver, spleen, kidney, lung, and tumor tissue were examined ex vivo from 6 to 72 h. The results showed enhanced blood circulation of the dendrimer conjugate compared to free DOX, specifically prolonged distribution up to 72 h (free DOX was cleared at 6 h) in tumor and kidney/liver and no dendrimer detection in the heart highlighting reduced potential for the adverse cardiac side effects traditionally associated with DOX (dilated cardiomyopathy leading to congestive heart failure; Chatterjee, Zhang, Honbo, & Karliner, 2010).

5. TRENDS IN CONJUGATION OF CHEMOTHERAPEUTICS TO DENDRONIZED SYSTEMS

5.1 Doxorubicin

One of the most common molecules conjugated to dendronized systems is doxorubicin, a molecule with potent anticancer activity against a wide variety of cancers but also possessing a range of adverse cardiac side effects (Carvalho et al., 2009). As a result of this activity, conjugation of doxorubicin (DOX) to dendronized systems that afford better control over the biodistribution and controlled release is crucial for practical applications.

She et al. synthesized G2-poly-L-lysine dendrons which were connected at the core to a heparin sulfate moiety (via "Click" chemistry), and doxorubicin was conjugated to the terminal ends of the dendron via acyl hydrazone bond (She et al., 2013). This amphiphilic dendron then assembled into 90 nm nanoparticles with a protected doxorubicin (9% drug loading by weight) core and heparin sulfate modified surface. In vitro drug release via the pH-sensitive acyl hydrazone bond was studied at pH 7.0 and 5.4. Rapid release of doxorubicin (84% of drug load) occurred up to 10 h at pH 5.4, while slow release up to 24% occurred at pH 7.0. In results discussed previously, the authors further evaluated these dendronized systems in vitro and observed significant tumor inhibition and reduced tumor volume compared to free DOX and saline control solutions over the course of 25 days (Fig. 11).

Wei et al. synthesized amphiphilic PAMAM dendrons by attaching two hydrophobic C18 alkyl chains to the dendron core ("Click" chemistry) and assembling them into DOX-loaded micelles (Wei et al., 2015). The resulting dendritic micelles were then evaluated for in vitro and in vivo therapeutic properties. Localization of the dendronized micelles was significantly enhanced in tumor tissue compared to free DOX (as measured by fluorescence microscopy) and the drug release of the DOX was found to be pH dependent (twice as much drug released at pH 5.4 vs pH 7.0). The authors hypothesized that this was due to the proton sponge effect, where increase electrostatic repulsion forces the release of more drug at acidic pH.

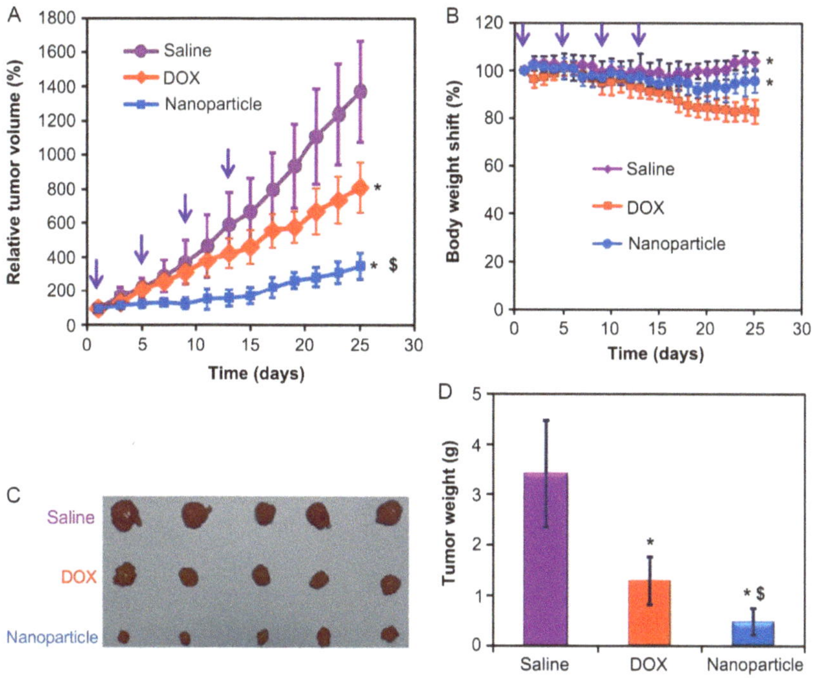

Fig. 11 In vivo tumor growth inhibition of nanoparticles. Comparison of the tumor inhibition effect of dendronized heparin–DOX conjugate-based nanoparticle (nanoparticle) vs free drug DOX and saline in the breast tumor model ($n=5$). The nanoparticle demonstrated significant tumor inhibition (*$P<0.001$, compared to saline; $^{\$}P<0.001$, compared to free drug DOX) (A). During the treatment, the mice-administrated nanoparticle showed no significant body lost compared to saline (*$P<0.001$, compared to free drug DOX) (B). At the end of this experiment, tumor tissues were collected from each sacrificed animal after 25 days treatment, photographed (C) and weighted (*$P<0.01$, compared to saline; $^{\$}P<0.05$, compared to free drug DOX) (D). *Reprinted with permission from She, W. C., Li, N., Luo, K., Guo, C. H., Wang, G., Geng, Y. Y., et al. (2013). Dendronized heparin-doxorubicin conjugate based nanoparticle as pH-responsive drug delivery system for cancer therapy.* Biomaterials, *34(9), 2252–2264 (Luo's group). Copyright 2013 Elsevier.*

Mice with xenograft tumors were injected with the dendronized micelle system and monitored over the course of 14 days. The tumor volume of xenograft-mice treated by the dendronized micelle system remained relatively consistent over the course of the 14 days, while mice treated with control (PBS and DOX-free dendronized system) or free DOX were observed to have a fourfold increase in tumor volume. The authors also observed decreased mouse mortality rate indicating that in addition to enhanced potency, this system had decreased toxicity, possibly due to lessened cardiac side effects.

5.2 Cisplatin

Cisplatin was incorporated by Li et al. into a stimuli-responsive nanoparticle for the treatment of a wide range of cancers (Li et al., 2016). PAMAM-cisplatin was grafted to a polycaprolactone polymer via 2-propionic-3-methylmaleic anhydride (PMA) for a pH-cleavable bond. This dendritic copolymer, PCL-PAMAM-PMA-cisplatin was assembled into the resulting nanoparticle (iCluster) with the addition of PCL–PEG to aid in prolonged circulation time and biocompatibility. The dendritic system showed longer circulation with low cisplatin release (15%) over 8 h in vitro followed by rapid cisplatin release after addition of ascorbic acid-containing buffer (similar to intracellular redox environment; 90% at 24 h). In vivo antitumor activity against mouse xenograft BxPC-3 cell lines (human pancreatic cancer) were performed. The authors observed a nearly sixfold increase in tumor cisplatin content (24 h postinjection) compared with free cisplatin in this dendronized system in addition to tumor growth suppression of 71%, 92%, and 107% for 1.5, 4.5, and 6.0 mg/kg dosages, respectively. The result of tumor reduction at 6.0 mg/kg was accompanied with an overall loss in body weight, potentially due to toxic side effects. Further extensive experiments were performed with metastatic cancer cell lines A549R (human lung cancer) and murine 4T1 (breast cancer) where this dendronized system showed additional promise in reducing tumor growth and prolonging survival of treated mice (100% of treated mice survived to 35 days at which point all control mice were dead).

In an effort to develop an efficient drug delivery platform to treat ovarian cancer, Kesavan et al. synthesized a G4-PAMAM-diglycolamic acid (DGA)-herceptin (also called trastuzumab, a targeting antibody) dendrimer to deliver a prodrug of cisplatin (Kesavan et al., 2015). Model cancer cell lines (SKOV-3 and PA-1) were treated with the G4-PAMAM-DGA-herceptin-cisplatin dendrimers and control groups were treated with dendrimers without targeting antibody or only free cisplatin. For the SKOV-3 cell line IC_{50} values of the dendronized system with antibody were observed to be 4.9 μg/mL at 48 h compared to 8.5 and 9.5 μg/mL for the dendrimer without antibody and free drug, respectively. Apoptosis induction for SKOV-3 and PA-1 cell lines treated with the dendrimer system was monitored for 18 h by flow cytometry. The results for SKOV-3 showed enhanced apoptotic activity for the dendrimers (in 73% of the cells) compared to untargeted dendrimer (56%) and free cisplatin (23%). The dendrimer, however, was less effective toward the PA-1 cancer cell line as the dendrimer systems showed lower apoptotic activity (19%) than free cisplatin (26%). In vivo experiments on mouse xenograft models with SKOV-3 tumors mirrored these results, showing significant tumor volume reduction over 2 weeks.

5.3 Taxols

Members of the taxane drug family such as paclitaxel (PTX) and docetaxel (DTX) are commonly conjugated to dendronized systems to aid in their delivery. Satsangi et al. demonstrated this by designing a two-part nanoplatform consisting of a PAMAM-GFLG-PTX dendrimers (selectively cleavable by cathepsin B within tumor cells) contained within folate-receptor targeting liposomes (Satsangi et al., 2015). With this system, the selective delivery of the PTX would be facilitated by folate-receptor targeting, and further enhanced by the tumor-selective cathepsin B cleavage to release PTX. The authors examined several breast cancer cell lines in vitro and found enhanced cytotoxic activity which correlated with the relative activity of cathepsin B or folate receptor expression in the different cell lines. The authors also performed in vivo experiments with the dendronized liposome nanoplatform in mouse xenograft models (MDA-MB-231 cells). All mice treated with the nanoplatform survived until the end of the experiment and showed no observable side effects, and tumor volume was reduced in the treated mice compared to liposomes containing only PTX and the saline control (Fig. 12).

In another study, delivery of DTX was achieved as a potential treatment for HER2+ types of breast cancer by conjugating a G4-PAMAM dendrimer to trastuzumab (antibody for HER2) and fluorescein isothiocyanate (FITC) for enhanced fluorescence imaging (Kulhari et al., 2016). The resulting G4-PAMAM-FITC-Ab dendrimers were loaded with DTX (encapsulation, concentration of 159.5 μg/mL) and evaluated in several in vitro studies. Drug release of the encapsulated DTX was initially rapid (55%) up to 10 h but then slowed down and reached 70% at 48 h, and showed a slower and more controlled release than free drug or PAMAM-DTX without Ab. In vitro cytotoxicity for the dendronized system was significantly higher than the control (empty PAMAM-Ab dendrimers) for two breast cancer cell lines, and the trastuzumab conjugate showed better selectivity toward the HER2+ over the HER2− cell line (Fig. 13). Additional experiments showed similar trends with uptake of the dendrimers as monitored by fluorescence microscopy and higher apoptosis (flow cytometry) in cancer cells treated with the dendronized system.

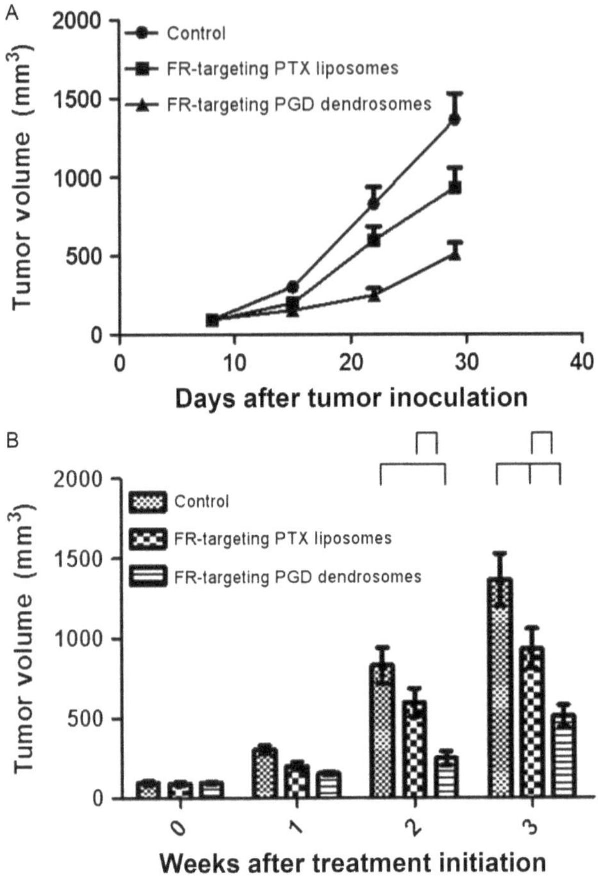

Fig. 12 Effect of FR-targeting PGD dendrosomes on human breast cancer (MDA-MB-231) xenograft tumors in mice. Therapy was administered every other day for three intraperitoneal injections total using the following treatments: control (saline), FR-targeting PTX liposomes, and FR-targeting PGD dendrosomes. Doses of 10 mg PTX/kg body weight were administered as FR-targeting PTX liposomes or FR-targeting PGD dendrosomes. Tumor volumes were measured 1, 2, and 3 weeks after therapy initiation and reported as mean ± standard error of the mean ($n = 7$). Statistically significant results are indicated by grouped lines ($P < 0.05$). Results are reported as two different representations (A) and (B) of the same data set. *Reprinted with permission from Satsangi, A., Roy, S. S., Satsangi, R. K., Tolcher, A. W., Vadlamudi, R. K., Goins, B., et al. (2015). Synthesis of a novel, sequentially active-targeted drug delivery nanoplatform for breast cancer therapy.* Biomaterials, 59, *88–101 (Satsangi's group). Copyright 2015 Elsevier.*

5.4 Camptothecin

Camptothecin (CPT) is known to be a potent topoisomerase inhibitor (Pommier, 2006) and has been a target for conjugation to dendronized systems. Zhou et al. designed a dendrimer where a single camptothecin

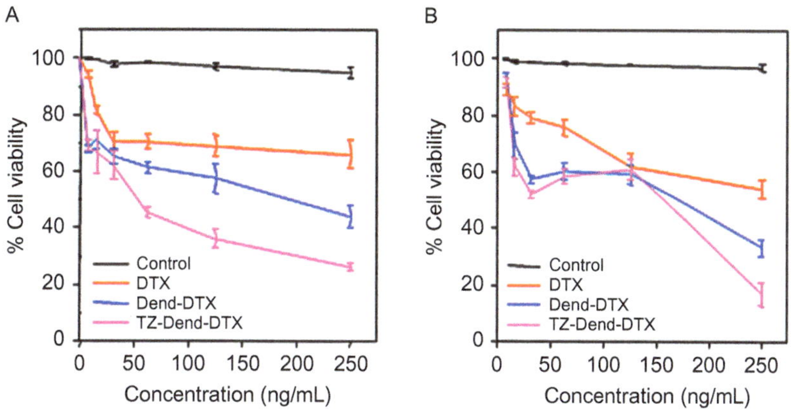

Fig. 13 Percentage cell viability of (A) MDA-MB-453 and (B) MDA-MB-231 human breast cancer cells treated with varying concentrations of docetaxel (DTX), DTX-loaded PAMAM dendrimers (Dend-DTX), and trastuzumab-conjugated Dend-DTX (TZ-Dend-DTX) (Mean $\pm$ SD; $n=3$). TZ-conjugated empty dendrimers were used as control. *Reprinted with permission from Kulhari, H., Pooja, D., Shrivastava, S., Kuncha, M., Naidu, V. G. M., Bansal, V., et al. (2016). Trastuzumab-grafted PAMAM dendrimers for the selective delivery of anticancer drugs to HER2-positive breast cancer.* Scientific Reports, *6, 23179 (Adams' group). Copyright 2016 Nature Publishing Group.*

molecule is the core, to which a G5-poly-L-lysine dendrimer was constructed (characterized by MALDI-TOF MS, NMR, HPLC, and UV/Vis) (Zhou et al., 2014). This synthesis is notable because rather than attach the drug with specific cleavable linkages, the dendrimer was observed to have a generation-dependent degradation and release profile. The fastest release (measured as 50% of the dendrimers) occurred with G1-CPT (0.42 h) and the slowest release with G2-CPT (21.3 h), however, successive generations higher than G2 resulted in more rapid release up to G6-CPT (5.3 h). Ultimately G5-poly-L-lysine dendrimer was used as the model system. The authors showed that hydrolysis of CPT was promoted by peripheral amine groups of the dendrimer, however, release occurred more slowly at acidic pH, which is not ideal for drug delivery. To solve this problem, the authors modified the terminal amine groups with a variety of β-carboxylic amides (βCA) which would promote hydrolysis of the core CPT molecule and enhance biocompatibility.

In vitro cytotoxicity experiments performed to determine the best G5-CPT-βCA derivative against SKOV-3 (ovarian) and MDA-MB-231 (breast) cancer cell lines found the dendrimer functionalized with 2,3-dimethyl-maleic anhydride to be ideal. This dendrimer (G5-CPT-DM) was then injected intravenously into mice containing SKOV-3 and MDA-MB-231 xenograft models. Compared to control (PBS), free CPT, and amine-protected O-[2-(2-methoxyethoxy)-ethyl]glycolic acid pentafluorophenol ester (OEG) dendrimers, the unmodified G5-CPT and G5-CPT-DM both resulted in significantly reduced tumor volumes and longer circulation time (Fig. 14).

5.5 Methotrexate

Methotrexate (MTX), an inhibitor of dihydrofolate reductase which prevents de novo synthesis of thymidine (Abolmaali, Tamaddon, & Dinarvand, 2013), is a common chemotherapeutic drug to treat a broad spectrum of cancers. Huang et al. synthesized a G5-PAMAM dendrimer conjugated to both MTX and glutamate urea (GLA, targeting small molecule for prostate-specific membrane antigen, PSMA) through similar copper-free "Click" reactions as discussed previously (Huang, Otis, Joice, Kotlyar, & Thomas, 2014). The GLA targeting ligand was attached through an amide bond while the drug was conjugated using ester or amide-linked variations. The G5-PAMAM dendrimer had on average 20 cyclooctyne groups to be functionalized with MTX or the GLA ligand. The reaction was carried out in one pot to form the conjugated dendrimer, with 3–8 GLA ligands and 8–10 MTX molecules (amide or ester bond) per dendrimer (characterized by NMR, MALDI-TOF, UHPLC/MS). In vitro cytotoxicity experiments were performed on two prostate cancer cell lines with and without expression of PSMA. The G5-PAMAM-MTX-GLA dendrimers showed strong selectivity for PSMA + cell lines and induced cytotoxicity, while PMSA − cell lines were unaffected.

In another example of effective delivery platforms for MTX, Sunoqrot et al. synthesized a hybrid nanoparticle system of G4-PAMAM dendrimers which were functionalized by on average four molecules each of rhodamine B, folic acid (FA), and MTX (characterized by ^{1}H NMR, HPLC, and UV/vis spectrophotometer) (Sunoqrot, Liu, Kim, & Hong, 2013). These dendrimers were then encapsulated within FITC-labeled PEG–PLA copolymers. Using fluorescence microscopy, the authors could observe selective uptake of the nanoparticles into cell containing the folate receptor and the nanoparticle system had enhanced penetration into tumor spheroids. In vitro cytotoxicity experiments on FR + cancer cell lines showed delayed growth inhibition compared to free MTX or only G4-PAMAM-MTX dendrimers, showing the targeted nanoparticles enhanced the selective and controlled release of MTX over the course of treatment.

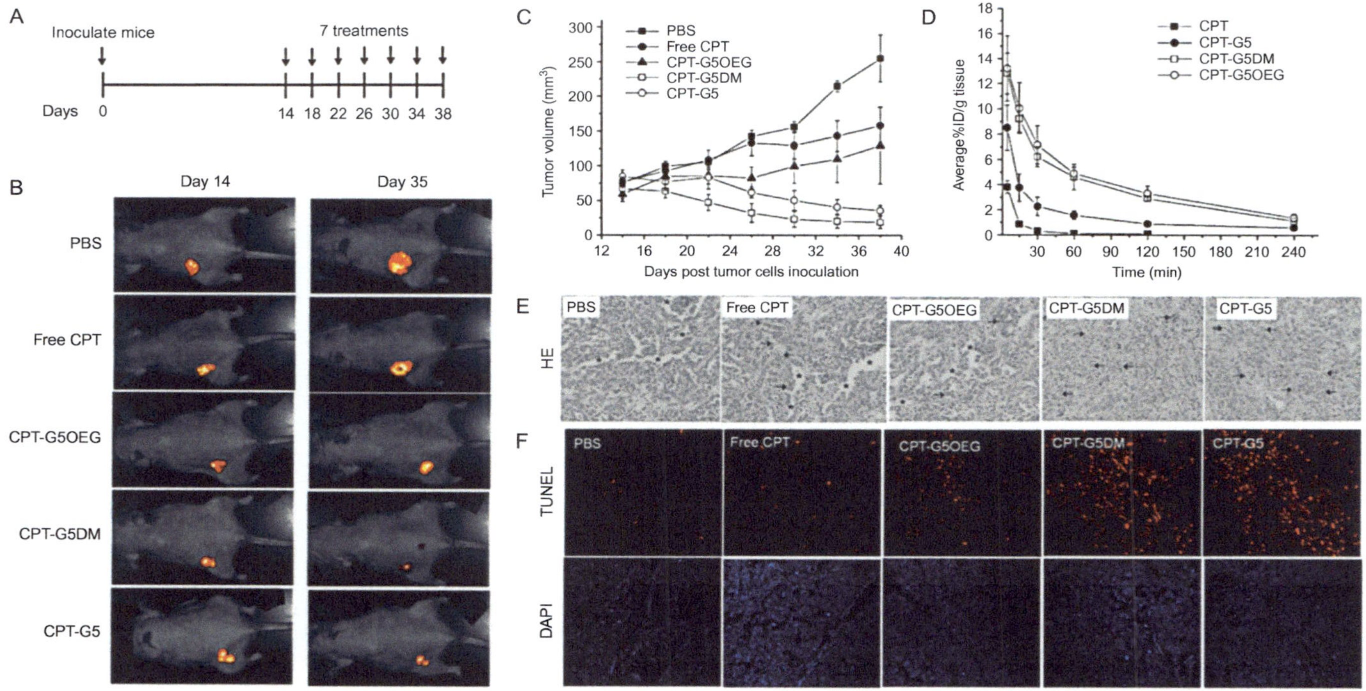

Fig. 14 (A) The treatment timeline and dosing schedule after the subcutaneous tumor inoculation in nude mice of MDA-MB-231-GFP breast cancer cells (10^6). The treatment was achieved by tail vein injection of CPT or the conjugates at 2.5 mg CPT equivalents per kg (each group $n=4$). (B) Representative in vivo Maestro fluorescent imaging of MD-MB-231-GFP tumors in nude mice before (day 14) and after six treatments (day 35). (C) The average tumor weight per mouse as a function of time of each group. (D) The blood clearance profiles after a single treatment. (E) Representative histological sections of the tumors stained with hematoxylin and eosin; blood vessels are marked by *; some typical pyknotic cells (highly condensed pyknotic nuclei) are indicated by *arrows*. (F) The TUNEL labeling for apoptosis in the tumor tissues was visualized by BrdUTP incorporation and detected using red fluorescence labeled anti-BrdU monoclonal antibody (*red*) and DAPI for nuclei (*blue*). *Reprinted with permission from Zhou, Z. X., Ma, X. P., Murphy, C. J., Jin, E. L., Sun, Q. H., Shen, Y. Q., et al. (2014). Molecularly precise dendrimer-drug conjugates with tunable drug release for cancer therapy.* Angewandte Chemie-International Edition, *53(41), 10949–10955 (Shen's group). Copyright 2014 Wiley.*

6. OUTLOOK AND CONCLUSIONS

The unique properties of dendronized systems discussed here make them promising tools for developing new and efficient methods to deliver chemotherapeutic drugs. While traditional cancer therapy is often limited by nonspecific cytotoxicity, in vitro and in vivo experiments with dendronized systems highlight a way to improve the therapeutic efficiency of drugs for the treatment of cancer.

Despite promising recent developments in areas such as biodegradable dendrimers, modifications to enhance biocompatibility, selectively cleavable drug conjugations, ligand-mediated targeting, and the potential for multifunctional properties, more advances are still needed to push this technology beyond in vitro and in vivo testing into the realm of clinical significance. Of primary interest are ways to control and understand the long-term in vivo effects to human health from exposure to dendrimers, and make the synthesis of dendronized nanoparticle systems feasible on larger industrial scales necessary for pharmaceutical development.

REFERENCES

Abolmaali, S. S., Tamaddon, A. M., & Dinarvand, R. (2013). A review of therapeutic challenges and achievements of methotrexate delivery systems for treatment of cancer and rheumatoid arthritis. *Cancer Chemotherapy and Pharmacology*, *71*(5), 1115–1130.

Acton, A. L., Fante, C., Flatley, B., Burattini, S., Hamley, I. W., Wang, Z. W., et al. (2013). Janus PEG-based dendrimers for use in combination therapy: Controlled multi-drug loading and sequential release. *Biomacromolecules*, *14*(2), 564–574.

Arseneault, M., Wafer, C., & Morin, J. F. (2015). Recent advances in click chemistry applied to dendrimer synthesis. *Molecules*, *20*(5), 9263–9294.

Binauld, S., & Stenzel, M. H. (2013). Acid-degradable polymers for drug delivery: A decade of innovation. *Chemical Communications*, *49*(21), 2082–2102.

Boas, U., & Heegaard, P. M. H. (2004). Dendrimers in drug research. *Chemical Society Reviews*, *33*(1), 43–63.

Brunetti, V., Bouchet, L. M., & Strumia, M. C. (2015). Nanoparticle-cored dendrimers: Functional hybrid nanocomposites as a new platform for drug delivery systems. *Nanoscale*, *7*(9), 3808–3816.

Carlmark, A., Hawker, C. J., Hult, A., & Malkoch, M. (2009). New methodologies in the construction of dendritic materials. *Chemical Society Reviews*, *38*(2), 352–362.

Carvalho, C., Santos, R. X., Cardoso, S., Correia, S., Oliveira, P. J., Santos, M. S., et al. (2009). Doxorubicin: The good, the bad and the ugly effect. *Current Medicinal Chemistry*, *16*(25), 3267–3285.

Chatterjee, K., Zhang, J. Q., Honbo, N., & Karliner, J. S. (2010). Doxorubicin cardiomyopathy. *Cardiology*, *115*(2), 155–162.

Chauhan, A. S., Jain, N. K., Diwan, P. V., & Khopade, A. J. (2004). Solubility enhancement of indomethacin with poly(amidoamine) dendrimers and targeting to inflammatory regions of arthritic rats. *Journal of Drug Targeting*, *12*(9–10), 575–583.

Chen, Q., Li, K. A., Wen, S. H., Liu, H., Peng, C., Cai, H. D., et al. (2013). Targeted CT/MR dual mode imaging of tumors using multifunctional dendrimer-entrapped gold nanoparticles. *Biomaterials, 34*(21), 5200–5209.

Cheng, Y. Y., Li, M. H., & Xu, T. W. (2008). Potential of poly(amidoamine) dendrimers as drug carriers of camptothecin based on encapsulation studies. *European Journal of Medicinal Chemistry, 43*(8), 1791–1795.

Cho, T. J., Zangmeister, R. A., MacCuspie, R. I., Patri, A. K., & Hackley, V. A. (2011). Newkome-type dendron-stabilized gold nanoparticles: Synthesis, reactivity, and stability. *Chemistry of Materials, 23*(10), 2665–2676.

Cline, E. N., Li, M. H., Choi, S. K., Herbstman, J. F., Kaul, N., Meyhofer, E., et al. (2013). Paclitaxel-conjugated PAMAM dendrimers adversely affect microtubule structure through two independent modes of action. *Biomacromolecules, 14*(3), 654–664.

Daniel, M.-C. (2012). Drug delivery carriers. In J. R. Lakowicz & J. Zhang (Eds.), *Emerging applications of colloidal noble metals in cancer nanomedicine* (pp. 54–67). London, UK: Future Medicine.

Daniel, M. C., Grow, M. E., Pan, H. M., Bednarek, M., Ghann, W. E., Zabetakis, K., et al. (2011). Gold nanoparticle-cored poly(propyleneimine) dendrimers as a new platform for multifunctional drug delivery systems. *New Journal of Chemistry, 35*(10), 2366–2374.

De Jesus, O. L. P., Ihre, H. R., Gagne, L., Frechet, J. M. J., & Szoka, F. C. (2002). Polyester dendritic systems for drug delivery applications: In vitro and in vivo evaluation. *Bioconjugate Chemistry, 13*(3), 453–461.

Deol, S., Weerasuriya, N., & Shon, Y. S. (2015). Stability, cytotoxicity and cell uptake of water-soluble dendron-conjugated gold nanoparticles with 3, 12 and 17 nm cores. *Journal of Materials Chemistry B, 3*(29), 6071–6080.

Dhanikula, R. S., Argaw, A., Bouchard, J. F., & Hildgen, P. (2008). Methotrexate loaded polyether-copolyester dendrimers for the treatment of gliomas: Enhanced efficacy and intratumoral transport capability. *Molecular Pharmaceutics, 5*(1), 105–116.

Esfand, R., & Tomalia, D. A. (2001). Poly(amidoamine) (PAMAM) dendrimers: From biomimicry to drug delivery and biomedical applications. *Drug Discovery Today, 6*(8), 427–436.

Feng, X. S., Pinaud, J., Chaikof, E. L., Taton, D., & Gnanou, Y. (2011). Sequential functionalization of janus-type dendrimer-like poly(ethylene oxide)s with camptothecin and folic acid. *Journal of Polymer Science Part A: Polymer Chemistry, 49*(13), 2839–2849.

Franc, G., & Kakkar, A. K. (2010). "Click" methodologies: Efficient, simple and greener routes to design dendrimers. *Chemical Society Reviews, 39*(5), 1536–1544.

Gerweck, L. E., & Seetharaman, K. (1996). Cellular pH gradient in tumor versus normal tissue: Potential exploitation for the treatment of cancer. *Cancer Research, 56*(6), 1194–1198.

Gillies, E. R., & Frechet, J. M. J. (2002). Designing macromolecules for therapeutic applications: Polyester dendrimer-poly(ethylene oxide) "bow-tie" hybrids with tunable molecular weight and architecture. *Journal of the American Chemical Society, 124*(47), 14137–14146.

Gillies, E. R., & Frechet, J. M. J. (2005). Dendrimers and dendritic polymers in drug delivery. *Drug Discovery Today, 10*(1), 35–43.

Guillaudeu, S. J., Fox, M. E., Haidar, Y. M., Dy, E. E., Szoka, F. C., & Frechet, J. M. J. (2008). PEGylated dendrimers with core functionality for biological applications. *Bioconjugate Chemistry, 19*(2), 461–469.

Hawker, C. J., & Frechet, J. M. J. (1990). Preparation of polymers with controlled molecular architecture. A new convergent approach to dendritic macromolecules. *Journal of the American Chemical Society, 112*(21), 7638–7647.

Hu, J. J., Cheng, Y. Y., Ma, Y. R., Wu, Q. L., & Xu, T. W. (2009). Host-guest chemistry and physicochemical properties of the dendrimer-mycophenolic acid complex. *Journal of Physical Chemistry B, 113*(1), 64–74.

Huang, B. H., Desai, A., Zong, H., Tang, S. Z., Leroueil, P., & Baker, J. R. (2011). Copper-free click conjugation of methotrexate to a PAMAM dendrimer platform. *Tetrahedron Letters*, *52*(13), 1411–1414.

Huang, B. H., Otis, J., Joice, M., Kotlyar, A., & Thomas, T. P. (2014). PSMA-targeted stably linked "dendrimer-glutamate urea-methotrexate" as a prostate cancer therapeutic. *Biomacromolecules*, *15*(3), 915–923.

Ihre, H., De Jesus, O. L. P., & Frechet, J. M. J. (2001). Fast and convenient divergent synthesis of aliphatic ester dendrimers by anhydride coupling. *Journal of the American Chemical Society*, *123*(25), 5908–5917.

Ihre, H. R., De Jesus, O. L. P., Szoka, F. C., & Frechet, J. M. J. (2002). Polyester dendritic systems for drug delivery applications: Design, synthesis, and characterization. *Bioconjugate Chemistry*, *13*(3), 443–452.

Jain, N. K., Tare, M. S., Mishra, V., & Tripathi, P. K. (2015). The development, characterization and in vivo anti-ovarian cancer activity of poly(propylene imine) (PPI)-antibody conjugates containing encapsulated paclitaxel. *Nanomedicine: Nanotechnology, Biology and Medicine*, *11*(1), 207–218.

Jevprasesphant, R., Penny, J., Jalal, R., Attwood, D., McKeown, N. B., & D'Emanuele, A. (2003). The influence of surface modification on the cytotoxicity of PAMAM dendrimers. *International Journal of Pharmaceutics*, *252*(1–2), 263–266.

Joralemon, M. J., O'Reilly, R. K., Matson, J. B., Nugent, A. K., Hawker, C. J., & Wooley, K. L. (2005). Dendrimers clicked together divergently. *Macromolecules*, *38*(13), 5436–5443.

Kaminskas, L. M., Boyd, B. J., Karellas, P., Krippner, G. Y., Lessene, R., Kelly, B., et al. (2008). The impact of molecular weight and PEG chain length on the systemic pharmacokinetics of PEGylated poly L-lysine dendrimers. *Molecular Pharmaceutics*, *5*(3), 449–463.

Kaminskas, L. M., McLeod, V. M., Porter, C. J. H., & Boyd, B. J. (2012). Association of chemotherapeutic drugs with dendrimer nanocarriers: An assessment of the merits of covalent conjugation compared to noncovalent encapsulation. *Molecular Pharmaceutics*, *9*(3), 355–373.

Kesavan, A., Ilaiyaraja, P., Beaula, W. S., Kumari, V. V., Lai, J. S., Arunkumar, C., et al. (2015). Tumor targeting using polyamidoamine dendrimer-cisplatin nanoparticles functionalized with diglycolamic acid and herceptin. *European Journal of Pharmaceutics and Biopharmaceutics*, *96*, 255–263.

Killops, K. L., Campos, L. M., & Hawker, C. J. (2008). Robust, efficient, and orthogonal synthesis of dendrimers via thiol-ene "Click" chemistry. *Journal of the American Chemical Society*, *130*(15), 5062.

Kim, S. T., Chompoosor, A., Yeh, Y. C., Agasti, S. S., Solfiell, D. J., & Rotello, V. M. (2012). Dendronized gold nanoparticles for siRNA delivery. *Small*, *8*(21), 3253–3256.

Kolb, H. C., Finn, M. G., & Sharpless, K. B. (2001). Click chemistry: Diverse chemical function from a few good reactions. *Angewandte Chemie, International Edition*, *40*(11), 2004.

Kono, K., Kojima, C., Hayashi, N., Nishisaka, E., Kiura, K., Wataral, S., et al. (2008). Preparation and cytotoxic activity of poly(ethylene glycol)-modified poly(amidoamine) dendrimers bearing adriamycin. *Biomaterials*, *29*(11), 1664–1675.

Krishna, T. R., Jain, S., Tatu, U. S., & Jayaraman, N. (2005). Synthesis and biological evaluation of 3-amino-propan-1-ol based poly(ether imine) dendrimers. *Tetrahedron*, *61*(17), 4281–4288.

Kulhari, H., Pooja, D., Shrivastava, S., Kuncha, M., Naidu, V. G. M., Bansal, V., et al. (2016). Trastuzumab-grafted PAMAM dendrimers for the selective delivery of anticancer drugs to HER2-positive breast cancer. *Scientific Reports*, *6*, 23179.

Kurtoglu, Y. E., Mishra, M. K., Kannan, S., & Kannan, R. M. (2010). Drug release characteristics of PAMAM dendrimer-drug conjugates with different linkers. *International Journal of Pharmaceutics*, *384*(1–2), 189–194.

Lee, C. C., Gillies, E. R., Fox, M. E., Guillaudeu, S. J., Frechet, J. M. J., Dy, E. E., et al. (2006). A single dose of doxorubicin-functionalized bow-tie dendrimer cures mice bearing C-26 colon carcinomas. *Proceedings of the National Academy of Sciences of the United States of America*, *103*(45), 16649–16654.

Lee, C. C., MacKay, J. A., Frechet, J. M. J., & Szoka, F. C. (2005). Designing dendrimers for biological applications. *Nature Biotechnology*, *23*(12), 1517–1526.

Leiro, V., Garcia, J. P., Tomas, H., & Pego, A. P. (2015). The present and the future of degradable dendrimers and derivatives in theranostics. *Bioconjugate Chemistry*, *26*(7), 1182–1197.

Li, H. J., Du, J. Z., Du, X. J., Xu, C. F., Sun, C. Y., Wang, H. X., et al. (2016). Stimuli-responsive clustered nanoparticles for improved tumor penetration and therapeutic efficacy. *Proceedings of the National Academy of Sciences of the United States of America*, *113*(15), 4164–4169.

Li, Y., He, H., Jia, X. R., Lu, W. L., Lou, J. N., & Wei, Y. (2012). A dual-targeting nanocarrier based on poly(amidoamine) dendrimers conjugated with transferrin and tamoxifen for treating brain gliomas. *Biomaterials*, *33*(15), 3899–3908.

Li, K. A., Wen, S. H., Larson, A. C., Shen, M. W., Zhang, Z. L., Chen, Q., et al. (2013). Multifunctional dendrimer-based nanoparticles for in vivo MR/CT dual-modal molecular imaging of breast cancer. *International Journal of Nanomedicine*, *8*, 2589–2600.

Li, N., Yi, Q. Y., Luo, K., Guo, C. H., Pan, D. Y., & Gu, Z. W. (2014). Amphiphilic peptide dendritic copolymer-doxorubicin nanoscale conjugate self-assembled to enzyme-responsive anti-cancer agent. *Biomaterials*, *35*(35), 9529–9545.

Maeda, H. (2015). Toward a full understanding of the EPR effect in primary and metastatic tumors as well as issues related to its heterogeneity. *Advanced Drug Delivery Reviews*, *91*, 3–6.

Majoros, I. J., Myc, A., Thomas, T., Mehta, C. B., & Baker, J. R. (2006). PAMAM dendrimer-based multifunctional conjugate for cancer therapy: Synthesis, characterization, and functionality. *Biomacromolecules*, 7(2), 572–579.

Majoros, I. J., Thomas, T. P., Mehta, C. B., & Baker, J. R. (2005). Poly(amidoamine) dendrimer-based multifunctional engineered nanodevice for cancer therapy. *Journal of Medicinal Chemistry*, *48*(19), 5892–5899.

Matsumura, Y., & Maeda, H. (1986). A new concept for macromolecular therapeutics in cancer chemotherapy: Mechanism of tumoritropic accumulation of proteins and the antitumor agent smancs. *Cancer Research*, *46*(12), 6387–6392.

McNerny, D. Q., Kukowska-Latallo, J. F., Mullen, D. G., Wallace, J. M., Desai, A. M., Shukla, R., et al. (2009). RGD Dendron bodies; synthetic avidity agents with defined and potentially interchangeable effector sites that can substitute for antibodies. *Bioconjugate Chemistry*, *20*(10), 1853–1859.

Medina, S. H., & El-Sayed, M. E. H. (2009). Dendrimers as carriers for delivery of chemotherapeutic agents. *Chemical Reviews*, *109*(7), 3141–3157.

Menjoge, A. R., Kannan, R. M., & Tomalia, D. A. (2010). Dendrimer-based drug and imaging conjugates: Design considerations for nanomedical applications. *Drug Discovery Today*, *15*(5–6), 171–185.

Mintzer, M. A., & Grinstaff, M. W. (2011). Biomedical applications of dendrimers: A tutorial. *Chemical Society Reviews*, *40*(1), 173–190.

Morgan, M. T., Nakanishi, Y., Kroll, D. J., Griset, A. P., Carnahan, M. A., Wathier, M., et al. (2006). Dendrimer-encapsulated camptothecins: Increased solubility, cellular uptake, and cellular retention affords enhanced anticancer activity in vitro. *Cancer Research*, *66*(24), 11913–11921.

Najlah, M., Freeman, S., Attwood, D., & D'Emanuele, A. (2006). Synthesis, characterization and stability of dendrimer prodrugs. *International Journal of Pharmaceutics*, *308*(1–2), 175–182.

Newkome, G. R., & Shreiner, C. D. (2008). Poly(amidoamine), polypropylenimine, and related dendrimers and dendrons possessing different 1→2 branching motifs: An overview of the divergent procedures. *Polymer, 49*(1), 1–173.

Newkome, G. R., Yao, Z. Q., Baker, G. R., & Gupta, V. K. (1985). Cascade molecules: A new approach to micelles. *Journal of Organic Chemistry, 50*(11), 2003–2004.

Nichols, J. W., & Bae, Y. H. (2014). EPR: Evidence and fallacy. *Journal of Controlled Release, 190*, 451–464.

Njikang, G. N., Gauthier, M., & Li, J. M. (2008). Sustained release properties of arborescent polystyrene-graft-poly(2-vinylpyridine) copolymers. *Polymer, 49*(25), 5474–5481.

Ornelas, C., Broichhagen, J., & Weck, M. (2010). Strain-promoted alkyne azide cycloaddition for the functionalization of poly(amide)-based dendrons and dendrimers. *Journal of the American Chemical Society, 132*(11), 3923–3931.

Ornelas, C., Pennell, R., Liebes, L. F., & Weck, M. (2011). Construction of a well-defined multifunctional dendrimer for theranostics. *Organic Letters, 13*(5), 976–979.

Pan, H. M., Grow, M. E., Wilson, O., & Daniel, M. C. (2013). A new poly(propylene imine) dendron as potential convenient building-block in the construction of multifunctional systems. *Tetrahedron, 69*(13), 2799–2806.

Papagiannaros, A., Dimas, K., Papaioannou, G. T., & Demetzos, C. (2005). Doxorubicin-PAMAM dendrimer complex attached to liposomes: Cytotoxic studies against human cancer cell lines. *International Journal of Pharmaceutics, 302*(1–2), 29–38.

Patri, A. K., Kukowska-Latallo, J. F., & Baker, J. R. (2005). Targeted drug delivery with dendrimers: Comparison of the release kinetics of covalently conjugated drug and non-covalent drug inclusion complex. *Advanced Drug Delivery Reviews, 57*(15), 2203–2214.

Patri, A. K., Majoros, I. J., & Baker, J. R. (2002). Dendritic polymer macromolecular carriers for drug delivery. *Current Opinion in Chemical Biology, 6*(4), 466–471.

Pommier, Y. (2006). Topoisomerase I inhibitors: Camptothecins and beyond. *Nature Reviews Cancer, 6*(10), 789–802.

Prajapati, R. N., Tekade, R. K., Gupta, U., Gajbhiye, V., & Jain, N. K. (2009). Dendimer-mediated solubilization, formulation development and in vitro-in vivo assessment of piroxicam. *Molecular Pharmaceutics, 6*(3), 940–950.

Quintana, A., Raczka, E., Piehler, L., Lee, I., Myc, A., Majoros, I., et al. (2002). Design and function of a dendrimer-based therapeutic nanodevice targeted to tumor cells through the folate receptor. *Pharmaceutical Research, 19*(9), 1310–1316.

Satsangi, A., Roy, S. S., Satsangi, R. K., Tolcher, A. W., Vadlamudi, R. K., Goins, B., et al. (2015). Synthesis of a novel, sequentially active-targeted drug delivery nanoplatform for breast cancer therapy. *Biomaterials, 59*, 88–101.

She, W. C., Li, N., Luo, K., Guo, C. H., Wang, G., Geng, Y. Y., et al. (2013). Dendronized heparin-doxorubicin conjugate based nanoparticle as pH-responsive drug delivery system for cancer therapy. *Biomaterials, 34*(9), 2252–2264.

Singh, P., Gupta, U., Asthana, A., & Jain, N. K. (2008). Folate and folate-PEG-PAMAM dendrimers: Synthesis, characterization, and targeted anticancer drug delivery potential in tumor bearing mice. *Bioconjugate Chemistry, 19*(11), 2239–2252.

Sun, L., Ma, X. F., Dong, C. M., Zhu, B. S., & Zhu, X. Y. (2012). NIR-responsive and lectin-binding doxorubicin-loaded nanomedicine from janus-type dendritic PAMAM amphiphiles. *Biomacromolecules, 13*(11), 3581–3591.

Sunoqrot, S., Liu, Y., Kim, D. H., & Hong, S. (2013). In vitro evaluation of dendrimer-polymer hybrid nanoparticles on their controlled cellular targeting kinetics. *Molecular Pharmaceutics, 10*(6), 2157–2166.

Teertstra, S. J., & Gauthier, M. (2004). Dendrigraft polymers: Macromolecular engineering on a mesoscopic scale. *Progress in Polymer Science, 29*(4), 277–327.

Teow, H. M., Zhou, Z. Y., Najlah, M., Yusof, S. R., Abbott, N. J., & D'Emanuele, A. (2013). Delivery of paclitaxel across cellular barriers using a dendrimer-based nanocarrier. *International Journal of Pharmaceutics*, *441*(1–2), 701–711.

Thomas, T. P., Majoros, I. J., Kotlyar, A., Kukowska-Latallo, J. F., Bielinska, A., Myc, A., et al. (2005). Targeting and inhibition of cell growth by an engineered dendritic nanodevice. *Journal of Medicinal Chemistry*, *48*(11), 3729–3735.

Tomalia, D. A., Baker, H., Dewald, J., Hall, M., Kallos, G., Martin, S., et al. (1985). A new class of polymers: Starburst-dendritic macromolecules. *Polymer Journal*, *17*(1), 117–132.

Tomalia, D. A., Naylor, A. M., & Goddard, W. A. (1990). Starburst dendrimers: Molecular-level control of size, shape, surface-chemistry, topology, and flexibility from atoms to macroscopic matter. *Angewandte Chemie International Edition*, *29*(2), 138–175.

Tsai, D. H., Cho, T. J., Elzey, S. R., Gigault, J. C., & Hackley, V. A. (2013). Quantitative analysis of dendron-conjugated cisplatin-complexed gold nanoparticles using scanning particle mobility mass spectrometry. *Nanoscale*, *5*(12), 5390–5395.

Waite, C. L., & Roth, C. M. (2009). PAMAM-RGD conjugates enhance siRNA delivery through a multicellular spheroid model of malignant glioma. *Bioconjugate Chemistry*, *20*(10), 1908–1916.

Wei, T., Chen, C., Liu, J., Liu, C., Posocco, P., Liu, X. X., et al. (2015). Anticancer drug nanomicelles formed by self-assembling amphiphilic dendrimer to combat cancer drug resistance. *Proceedings of the National Academy of Sciences of the United States of America*, *112*(10), 2978–2983.

Wen, S. H., Li, K. G., Cai, H. D., Chen, Q., Shen, M. W., Huang, Y. P., et al. (2013). Multifunctional dendrimer-entrapped gold nanoparticles for dual mode CT/MR imaging applications. *Biomaterials*, *34*(5), 1570–1580.

Wooley, K. L., Hawker, C. J., & Frechet, J. M. J. (1991). Hyperbranched macromolecules via a novel double-stage convergent growth approach. *Journal of the American Chemical Society*, *113*(11), 4252–4261.

Wu, P., Feldman, A. K., Nugent, A. K., Hawker, C. J., Scheel, A., Voit, B., et al. (2004). Efficiency and fidelity in a click-chemistry route to triazole dendrimers by the copper (I)-catalyzed ligation of azides and alkynes. *Angewandte Chemie, International Edition*, *43*(30), 3928–3932.

Wu, L. P., Ficker, M., Christensen, J. B., Trohopoulos, P. N., & Moghimi, S. M. (2015). Dendrimers in medicine: Therapeutic concepts and pharmaceutical challenges. *Bioconjugate Chemistry*, *26*(7), 1198–1211.

Wu, P., Malkoch, M., Hunt, J. N., Vestberg, R., Kaltgrad, E., Finn, M. G., et al. (2005). Multivalent, bifunctional dendrimers prepared by click chemistry. *Chemical Communications*, *46*, 5775–5777.

Wurm, F., & Frey, H. (2011). Linear-dendritic block copolymers: The state of the art and exciting perspectives. *Progress in Polymer Science*, *36*(1), 1–52.

Yoon, K., Goyal, P., & Weck, M. (2007). Monofunctionalization of dendrimers with use of microwave-assisted 1,3-dipolar cycloadditions. *Organic Letters*, *9*(11), 2051–2054.

Zhang, C. Y., Pan, D. Y., Luo, K., Li, N., Guo, C. H., Zheng, X. L., et al. (2014). Dendrimer-doxorubicin conjugate as enzyme-sensitive and polymeric nanoscale drug delivery vehicle for ovarian cancer therapy. *Polymer Chemistry*, *5*(18), 5227–5235.

Zhang, L. H., Zhu, S. J., Qian, L. L., Pei, Y. Y., Qiu, Y. M., & Jiang, Y. Y. (2011). RGD-modified PEG-PAMAM-DOX conjugates: In vitro and in vivo studies for glioma. *European Journal of Pharmaceutics and Biopharmaceutics*, *79*(2), 232–240.

Zhou, Z. Y., D'Emanuele, A., & Attwood, D. (2013). Solubility enhancement of paclitaxel using a linear-dendritic block copolymer. *International Journal of Pharmaceutics*, *452*(1–2), 173–179.

Zhou, Z. X., Ma, X. P., Murphy, C. J., Jin, E. L., Sun, Q. H., Shen, Y. Q., et al. (2014). Molecularly precise dendrimer-drug conjugates with tunable drug release for cancer therapy. *Angewandte Chemie, International Edition*, *53*(41), 10949–10955.
Zhu, S. J., Qian, L. L., Hong, M. H., Zhang, L. H., Pei, Y. Y., & Jiang, Y. Y. (2011). RGD-modified PEG-PAMAM-DOX conjugate: In vitro and in vivo targeting to both tumor neovascular endothelial cells and tumor cells. *Advanced Materials*, *23*(12), H84–H89.

CHAPTER FIVE

Thermosensitive Liposomes for Image-Guided Drug Delivery

Dieter Haemmerich*,[1], Anjan Motamarry*,†
*Department of Pediatrics, Medical University of South Carolina, Charleston, SC, United States
†Department of Drug Discovery and Biomedical Sciences, Medical University of South Carolina, Charleston, SC, United States
[1]Corresponding author: e-mail address: haemmer@musc.edu

Contents

Abstract

Liposomes have been employed as cancer therapy clinically since the 1990s, with the primary benefit of reduced toxicity but no appreciable efficacy improvement. Thermosensitive liposomes (TSLs) are specifically formulated such that they release the encapsulated drug when exposed to hyperthermic temperatures in the fever range (~40–42°C) and have been investigated as cancer therapy for several decades, with first clinical trials initiated in the last decade. Combined with localized hyperthermia, TSLs allow precise drug delivery to a targeted region. Typically, the targeted tissue is exposed to localized hyperthermia facilitated by an image-guided hyperthermia device. Thus, TSLs enable image-guided drug delivery where drug is delivered to a tissue region identified by medical imaging. Recent TSL formulations are based on the more recent paradigm of intravascular triggered release, where drug is released rapidly (within seconds) while TSLs pass through the vasculature of the heated tissue region. The drug released within the blood then extravasates and is taken up by cancer cells. These TSLs enable up to 20–30 times higher tumor drug uptake compared to infusion of unencapsulated drug, and the dose locally delivered to the heated region can be modulated based on heating duration. This chapter reviews various TSL formulations, the different anticancer agents that have been encapsulated, as well as targeted cancer types. Further, the various hyperthermia

Advances in Cancer Research, Volume 139
ISSN 0065-230X
https://doi.org/10.1016/bs.acr.2018.04.004

devices that have been used for image-guided hyperthermia are reviewed, focusing on those that have been employed in human patients.

ABBREVIATIONS

CF carboxyfluorescein
Cho cholesterol
CT computed tomography
Dox doxorubicin
DPPC dipalmitoyl phosphatidylcholine
DPPG$_2$ 1,2-dipalmitoyl-*sn*-glycero-3-phosphoglyceroglycerol
DSPC distearoyl phosphatidylcholine
EPR enhanced permeability and retention
HIFU high-intensity focused ultrasound
LTSL lysolipid temperature-sensitive liposomes
MRgHIFU magnetic resonance-guided high-intensity focused ultrasound
MRI magnetic resonance imaging
PEG polyethylene glycol
RFA radiofrequency ablation
TSL thermosensitive liposomes
TTSL traditional thermosensitive liposomes

1. INTRODUCTION

Liposomes are nanoparticle drug delivery systems that have been in clinical use for targeted chemotherapy delivery since the 1990s (Barenholz, 2012). Encapsulation of a drug (e.g., chemotherapy agent) in liposomes prevents drug uptake by noncancer tissues and takes advantage of the enhanced permeability and retention (EPR) effect to facilitate preferential accumulation of liposomal drug within the tumor interstitium (Fig. 1B) (Kneidl, Peller, Winter, Lindner, & Hossann, 2014; Kunjachan, Ehling, Storm, Kiessling, & Lammers, 2015). Since this tumor accumulation based on passive targeting via the EPR effect requires typically ~24–48 h, adequate liposome stability is essential. For example, liposomes have to evade the immune and reticuloendothelial systems, which is achieved by incorporation of polyethylene glycol (PEG) (Mumtaz, Ghosh, & Bachhawat, 1991). This PEG incorporation enhances the liposomal plasma half-life from a few hours to several days, and these types of liposomes are usually termed "stealth liposomes" due to their ability to remain undetected by the immune and reticuloendothelial systems. Addition of antibodies or targeting molecules to liposomes may help actively target tumors instead of depending on solely EPR effect alone (Fig. 1C and D), but such targeting is not employed in current clinical liposome formulations. While stealth liposomes are in clinical use as cancer therapy, their primary benefit

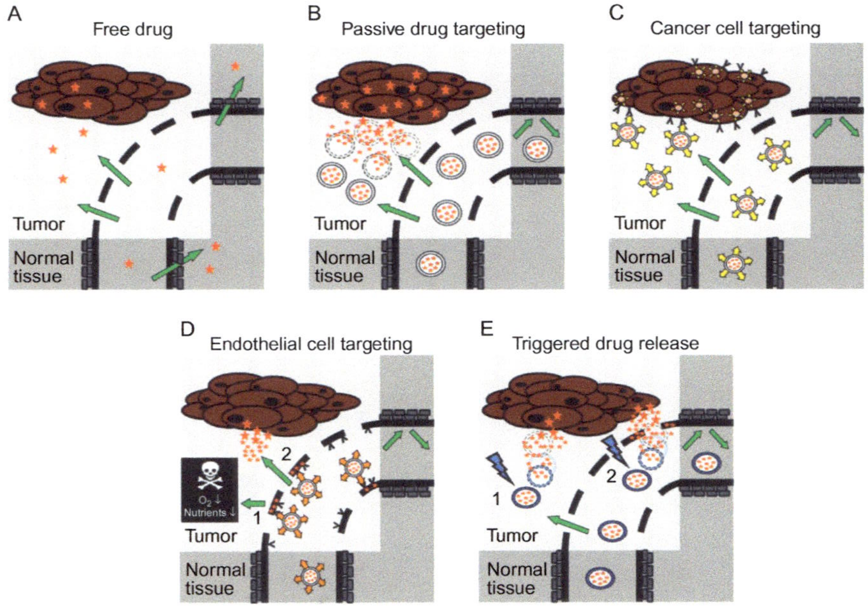

Fig. 1 Drug-targeting strategies. (A) Conventional therapy or free drug infusion. (B) Passive targeting by liposomes utilizing EPR effect. (C) Active targeting of liposomes labeled with tumor cell-specific antibody. (D) Active targeting of liposomes with endothelial cell-specific antibody. (E) Triggered drug release either (1) within the tumor interstitium or (2) within vasculature. TSLs fall under the category (E). *Reproduced with permission from Kunjachan, S., Ehling, J., Storm, G., Kiessling, F., & Lammers, T. (2015). Noninvasive Imaging of nanomedicines and nanotheranostics: Principles, Progress, and Prospects.* Chemical Reviews, 115*(19), 10907–10937. https://doi.org/10.1021/cr500314d. Copyright (2015) American Chemical Society.*

has been the reduction of systemic toxicities with no appreciable impact on efficacy (Park, 2013). Limitations of stealth liposomes are (1) heterogeneity in tumor vascular permeability (within a tumor and between tumors), making tumor liposome uptake variable; (2) limited bioavailability, as drug may remain encapsulated even if taken up by cancer cells; liposomes may be degraded within the endosomes or lysosomal compartments within the cells.

To improve targeting of liposomes, triggered release liposomes have been designed. These types of liposomes release the encapsulated cargo upon externally applied stimuli (e.g., heat, laser) or internally inherent stimuli (e.g., pH). Here, we will discuss a specific class of triggered liposomes termed thermosensitive liposomes (TSLs), where release is elicited by exposure to hyperthermic temperatures in the fever range, typically above 40°C. Similar to stealth liposomes, TSLs are typically administered as intravenous infusion and have extended plasma half-life compared to unencapsulated drug.

TSL may be employed based on either of two delivery strategies: (1) Extravascular triggered delivery, where TSLs passively accumulate within the tumor interstitium followed by triggered release, or (2) intravascular triggered delivery, where release occurs while TSLs pass through the vasculature of the heated tissue region (Fig. 1E). The drug then penetrates the surrounding tissue interstitium and is ultimately taken up by the cells. In animal studies, TSLs have been shown to deliver up to 20–30 times higher drug concentration to target tissues compared to free (i.e., unencapsulated) drug (Kong et al., 2000). For comparison, stealth liposomes can achieve 5–10 times higher tumor concentration than free drug, but most of the delivered drug is not bioavailable since it remains encapsulated.

Drug Formulation	Improvement of Tumor Drug Delivery	Comments
Traditional chemotherapy by free drug infusion (e.g., doxorubicin)	×1 (free drug used as reference)	Toxic side effects, often insufficient delivery to the target tumor tissue
Stealth liposomes (e.g., Doxil©)	×5–10 (Park, 2013)	Reduced toxicities. Tumor drug uptake depends on tumor type. Drug not readily bioavailable
Thermosensitive liposomes (TSLs)	×20–30 (Kong et al., 2000)	Following release, drug is bioavailable similar to free drug infusion. Reduced toxicities similar to stealth liposomes. Local dose can be controlled by hyperthermia duration

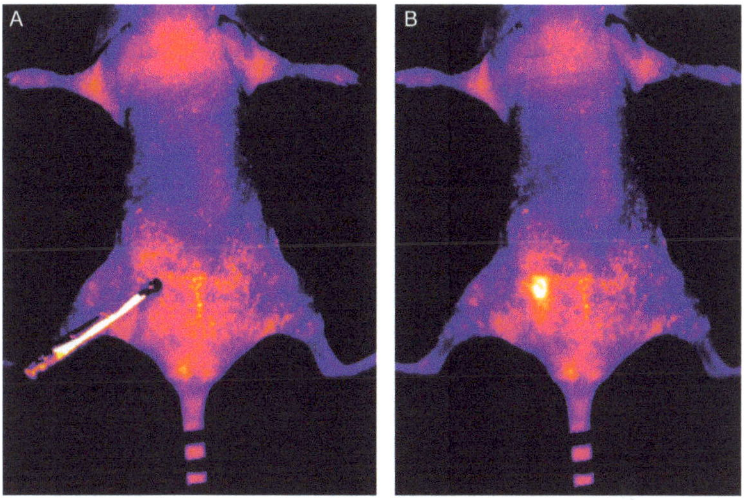

Fig. 2 Targeted drug delivery with TSL. (A) Hyperthermia is locally applied to subcutaneous tissues by a heated probe placed in contact with skin in mice, following intravenous administration of TSL encapsulated doxorubicin. (B) Fluorescent images demonstrate highly localized delivery of the fluorescent chemotherapy agent doxorubicin (image was obtained immediately following 10 min hyperthermia). *© 2018 IEEE. Reprinted, with permission, from Asemani, D., & Haemmerich, D. (2018). A unified mathematical model for nano-liposomal drug delivery to solid tumors. IEEE Transactions on NanoBioscience, 17, 3–11. https://doi.org/10.1109/TNB.2017.2783889.*

Contrary to many other nanoparticle delivery systems, TSL can facilitate image-guided drug delivery, where drug is delivered to a tissue region defined by medical imaging. This approach is possible since drug is released from TSL and taken up by tissue only in tissue regions where the required temperature for release is achieved (>40°C). That means with the combination of TSL and a heating technology that allows heat exposure of a targeted tissue region with high spatial accuracy, spatial targeting of delivery with mm accuracy may be feasible (Fig. 2). This also means that the location of tissue regions where cancer is present (or where the presence is suspected) has to be known; i.e., TSLs are likely not ideal for metastatic disease where cancer cells have spread to multiple organ sites. Several technologies are available that allow local, image-guided tissue heating, and selection of an appropriate hyperthermia technology is important for providing the optimal delivery with TSL.

2. EVOLUTION OF TSL FORMULATIONS

Evolution of present day "fast-release TSLs" occurred over several decades since the first study by Yatvin et al. introduced the concept of TSL for treating bacterial infections (Yatvin, Weinstein, Dennis, & Blumenthal, 1978). Since then, TSLs have overwhelmingly been investigated for encapsulating and delivering chemotherapeutic drugs. The very first TSL formulations were composed of just two lipids, namely, dipalmitoyl phosphatidylcholine (DPPC) and distearoyl phosphatidylcholine (DSPC). These lipids were selected due to their very low melting point, also called transition temperature (T_m). DPPC has a T_m of 41°C at which its impermeable solid crystalline structure turns into a highly permeable liquid structure. As liposomes made of single lipids were not sufficiently stable at body temperature, combinations of lipids were used. However, there were significant problems with these first formulations—later termed "traditional thermosensitive liposomes (TTSLs)"—which have been summarized by Dewhirst, Landon, Hofmann, and Stauffer (2013):

(a) TTSLs were eliminated in less than an hour by the reticuloendothelial and immune systems.
(b) Rate of release of the drug from the TTSLs was slow.
(c) Drug release from TTSLs occurred at the higher end of hyperthermia range (43–45°C) where hyperthermia can induce shutdown of perfusion.
(d) TTSLs were very leaky at physiological body temperature of 37°C.

The addition of PEG (Mumtaz et al., 1991; Papahadjopoulos et al., 1991) and optimization of the liposome size (Magin, Hunter, Niesman, & Bark, 1986) seemed to overcome the first problem. Several workers experimented with adding cholesterol to TTSLs as component. While adding cholesterol can improve TSL stability (Sadeghi et al., 2017), it may also delay drug release during hyperthermia (Gaber, Hong, Huang, & Papahadjopoulos, 1995). Rapid release turned out to be important when TSLs are employed based on intravascular triggered release, as the blood carrying the TSL pass through the heated region only for a few seconds (Burke et al., 2017). Needham and Dewhirst proposed incorporation of lysolipids into TSLs (Needham, Anyarambhatla, Kong, & Dewhirst, 2000). Lysolipids are lipid acyl derivates that are more hydrophilic than conventional lipids. The incorporation of lysolipids and PEG seemed to achieve rapid release while obtaining adequate stability. Hence these TSLs have become known as lysolipid TSLs (LTSLs), and this formulation achieved complete release within 2 s at mild hyperthermia (~42°C) (Burke et al., 2017). The rapid release of drug from TSLs consisting of multiple lipids is facilitated by inhomogenous membrane microstructures that develop between lipid interfaces (Lu & Ten Hagen, 2017). LTSL loaded with doxorubicin is the only formulation that has made it to human clinical trials. Yet despite of all these advances, there is still a considerable problem with leakiness of TSLs at body temperature that contributes toward a limited initial plasma half-life of ~1 h in human patients (Wood et al., 2012) and ~30 min in canines (Hauck et al., 2006). Lindner et al. proposed TSLs with 1,2-dipalmitoyl-*sn*-glycero-3-phosphoglyceroglycerol ($DPPG_2$), which showed improved plasma half-life while exhibiting quick release during hyperthermia. While initial studies encapsulating the fluorescent dye carboxyfluorescein showed extended plasma half-life of 5–10 h in small animals (Lindner et al., 2004), and loading of doxorubicin showed good in vitro stability at body temperature (Hossann et al., 2007), recent studies suggest in vivo stability similar to Needham's LTSL formulation, with initial plasma half-life of around 40 min in felines (Zimmermann et al., 2017). Several workers have also reported TSLs made with polymers and temperature-sensitive hydrogels (Hotta et al., 2016; Kono, 2001). While such polymers have been getting increased attention, they are not biodegradable which may pose risks related to long-term toxicities.

The TSL release kinetics depend both on the liposomal formulation and on the encapsulated compound (Fig. 3) (Burke et al., 2017), i.e., TSLs need to be individually selected and optimized depending on the drug to be delivered.

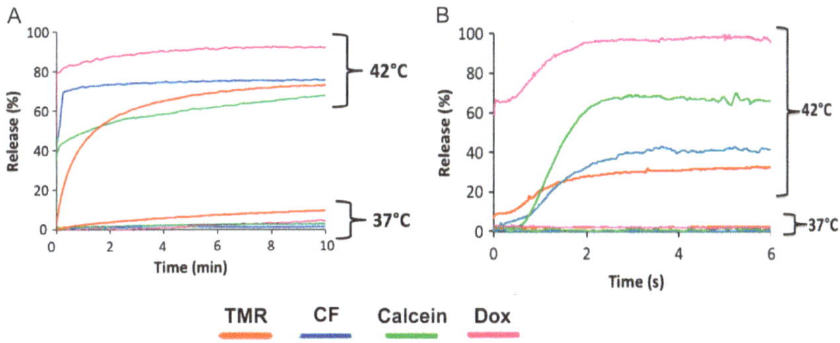

Fig. 3 Release of four compounds from a MPSC-TSL formulation. Release is plotted at 37°C and 42°C (A) during the first 10 min, and (B) during the first 6 s for three fluorescent dyes (tetramethylrhodamine (TMR), carboxyfluorescein (CF), and Calcein) and for the drug doxorubicin (Dox). *Copyright Taylor & Francis. Reproduced with permission from Burke, C., Dreher, M. R., Negussie, A. H., Mikhail, A. S., Yarmolenko, P., Patel, A., et al. (2017). Drug release kinetics of temperature sensitive liposomes measured at high-temporal resolution with a millifluidic device.* International Journal of Hyperthermia, *1–9. https://doi.org/10.1080/02656736.2017.1412504.*

2.1 Targeted TSL

Several workers have added additional targeting molecules for the active targeting of tumors. The various targeting molecules that have been tried in combination with TSLs are antibodies, aptamers, and peptides and have been summarized in the table below:

Type	Targeting Molecule	Target	Payload	References
Antibody	Anti-H2Kk	H2Kk	Carboxyfluorescein	Sullivan and Huang (1985)
Antibody	Anti-OV-TL3	OA3	Doxorubicin	Vingerhoeds et al. (1996)
Antibody	Anti-EGFR	EGFR	Calcein, doxorubicin	Haeri, Zalba, Ten Hagen, Dadashzadeh, and Koning (2016, pp. 657–669)
Antibody	Anti-Her-2	Her-2	Calcein	Kullberg, Mann, and Owens (2009)
Peptide	VPGVG	Intracellular delivery	Doxorubicin	Na, Lee, Jung, Hyun, and Shin (2012)
Peptide	CREKA	Clotted plasma proteins in tumor vessels	Doxorubicin	Wang et al. (2015)
Peptide	RGDf[N-Met]C	Angiogenic endothelial cells	Doxorubicin	Dicheva et al. (2015)
Peptide	CCRGDKGPDC	$\alpha\nu\beta3$-positive cells	Doxorubicin	Deng et al. (2016)
Aptamer	AS1411	Nucleolin receptor	Contrast agent	Zhang et al. (2015)
Aptamer	AS1411 & S2.2	Nucleolin receptor	Doxorubicin	Zhao et al. (2017)

Several antibodies including H2Kk (Sullivan & Huang, 1985), OV-TL3 (Vingerhoeds et al., 1996), Her-2 (Kullberg et al., 2009), and EGFR (Haeri et al., 2016) have been attached to TSLs to deliver a variety of fluorescent molecules and therapeutic agents such as doxorubicin. While several of the initial studies only evaluated the in vitro efficacy, Vingerhoeds et al. showed that antibody-labeled TSLs were not effective in vivo (Vingerhoeds et al., 1996) which was attributed to the leakiness of the liposomes. Newer formulations of TSLs have been comparatively more stable than the formulation used by them. Recently Haeri et al. evaluated a newer formulation and labeled them with anti-EGFR antibody in vitro. They found that the addition of antibody to the TSLs increases the intracellular quantity of the drug (Haeri et al., 2016).

Similarly, several peptides have also been attached to the TSLs. Several workers have shown that using a peptide as homing molecule does increase the intracellular delivery in vitro (Dicheva et al., 2015) (Na et al., 2012) (Wang et al., 2015). Recently, aptamers as targeting molecules for TSLs have been reported. Zhang et al. showed that contrast agent could be loaded to aptamer labeled TSLs for efficient magnetic resonance imaging (MRI) (Zhang et al., 2015). Zhao et al. used multimodal photodynamic therapy with gold-labeled TSLs loaded with doxorubicin (Zhao et al., 2017). Although targeted TSLs have been shown to increase intracellular drug delivery into in vitro, this has yet to be shown in vivo. Since TSLs targeted toward cancer cells require extravasation, targeted TSLs likely do not provide benefit when employing TSL based on intravascular triggered delivery.

3. TSL DELIVERY KINETICS

As mentioned earlier, TSL can be employed based on intra- or extravascular triggered release strategy. Extravascular triggered release requires tumor accumulation of liposomes (e.g., based on EPR effect) before initiating release of the accumulated TSL. During intravascular triggered delivery, the TSL in the systemic plasma pool provide a continuous supply of nonbioavailable drug that becomes bioavailable when entering the vasculature of the heated tissue region. With rapid-release TSL, this paradigm is conceptually similar to the direct infusion of unencapsulated drug into the tumor vasculature for prolonged duration. Following release within the vasculature, the drug extravasates into the interstitium and is then taken

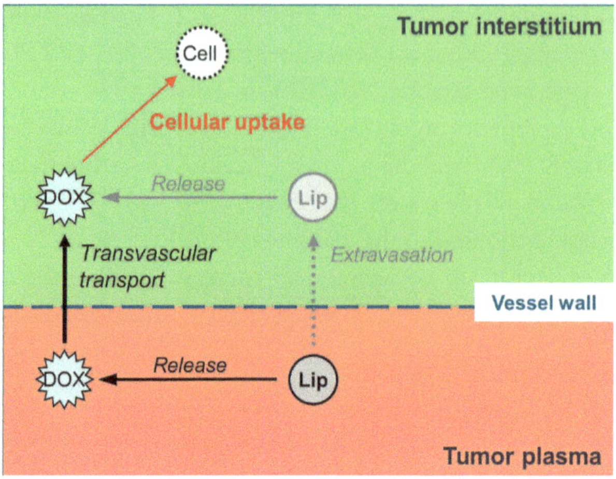

Fig. 4 TSL can be employed by one of two delivery strategies: (1) Extravascular triggered delivery, which requires TSL extravasation followed by triggered release (*gray arrows*); or (2) Intravascular triggered delivery, where release is triggered within the vasculature. The released drug is then take up by interstitium, and cancer cells (*black arrows*). *Reproduced with permission. Originally published in Gasselhuber, A., Dreher, M. R., Rattay, F., Wood, B. J., & Haemmerich, D. (2012). Comparison of conventional chemotherapy, stealth liposomes and temperature-sensitive liposomes in a mathematical model.* PLoS One, *7(10), e47453. https://doi.org/10.1371/journal.pone.0047453 under CC BY 3.0 license.*

up by the cells (Fig. 4). These two delivery approaches have been compared in computer simulations (Fig. 5) (Gasselhuber, Dreher, Rattay, Wood, & Haemmerich, 2012).

While for intravascular triggered delivery rapid drug release from TSL (within seconds) is important, the optimal release rate for extravascular triggered delivery is in the range of hours (El-Kareh & Secomb, 2000; Gasselhuber, Dreher, Rattay, et al., 2012). The computer simulation also suggests that intravascular triggered delivery is more efficacious than extravascular triggered delivery (Fig. 5), and this is supported by an in vivo study in mice that compared the two approaches (Li et al., 2013).

As suggested by Fig. 5, during intravascular triggered delivery, the hyperthermia duration impacts amount of drug released and delivered to cancer cells. As long as hyperthermia is applied, liposomal encapsulated drug enters the vasculature of the heated target area and is released. The enhanced delivery with longer heating duration has recently been confirmed by in vivo studies in large animals (Rossmann, McCrackin, Armeson, & Haemmerich, 2017). In addition to hyperthermia duration,

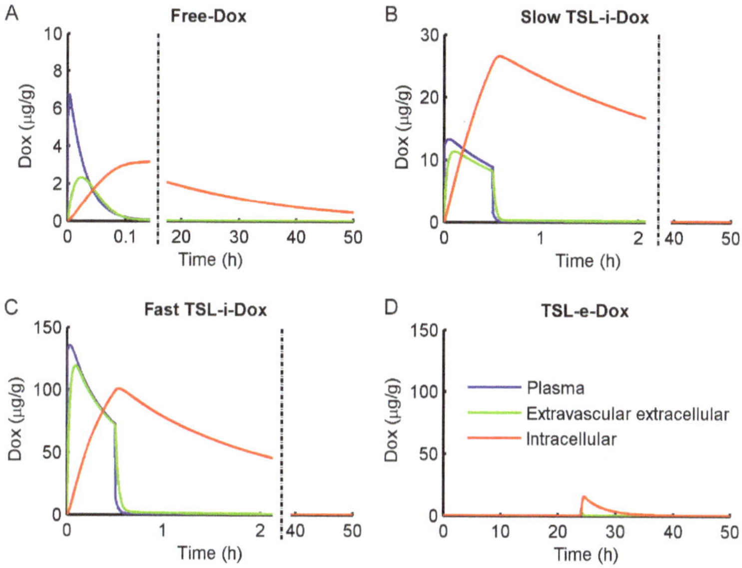

Fig. 5 Plasma, EES, and intracellular tumor concentration time courses of bioavailable doxorubicin for (A) unencapsulated drug administration (bolus injection), (B) intravascular triggered delivery with slow-release TSL (complete release within ~2–3 min), (C) intravascular triggered delivery with fast-release TSL (complete release within ~3–4 s), and (D) extravascular triggered delivery, TSL with optimized release rate (complete release within ~1 h). For (D), 24 h extravasation was considered with subsequent triggered release from TSL in interstitium. Note that EES and intracellular concentrations are relative to EES and cell volume, i.e., would need to be multiplied by appropriate volume fractions for conversion to tissue concentration. *Reproduced with permission. Originally published in Gasselhuber, A., Dreher, M. R., Rattay, F., Wood, B. J., & Haemmerich, D. (2012). Comparison of conventional chemotherapy, stealth liposomes and temperature-sensitive liposomes in a mathematical model.* PLoS One, *7(10), e47453. https://doi.org/10.1371/journal.pone.0047453 under CC BY 3.0 license.*

the plasma concentration of encapsulated drug impacts amount available for delivery, as it represents the drug amount available for intravascular release. The plasma area under the curve (AUC) of encapsulated drug calculated during heating is thus representative of the total amount of drug released during heating. A recent computer modeling study suggests that this plasma AUC of encapsulated drug calculated during heating is predictive of amount of drug delivered to the target tissue (Fig. 6), and this relationship was recently confirmed in vivo in canines (Bredlau et al., 2018). These results suggest (1) the importance of applying hyperthermia when plasma concentration of encapsulated drug is highest,

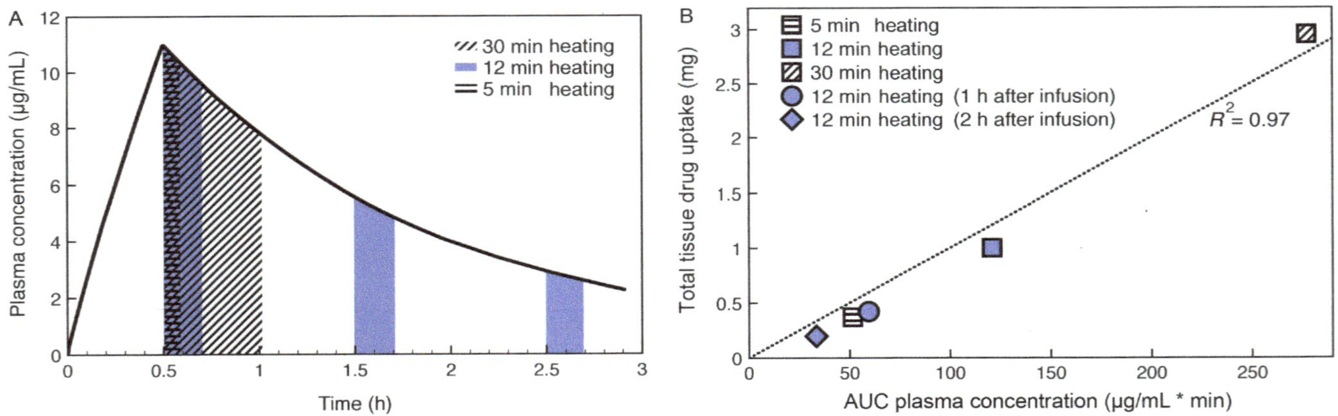

Fig. 6 Plasma AUC of encapsulated drug predicts tissue uptake (computer simulation results): (A) Systemic plasma concentration of TSL-Dox (i.e., encapsulated Dox). Areas under the curve (AUCs) are *colored/shaded* for various heating cycles that were compared: heating was initiated for three durations (5, 12, and 30 min duration) immediately following a 30-min TSL-Dox infusion, or following 1 and 2 h after infusion (12 min heating duration). (B) Systemic plasma AUC of TSL-Dox calculated during heating correlates with total amount of doxorubicin delivered to the target tissue ($R^2 = 0.97$). *Reproduced with permission. Originally published in Rossmann, C., McCrackin, M. A., Armeson, K. E., & Haemmerich, D. (2017). Temperature sensitive liposomes combined with thermal ablation: Effects of duration and timing of heating in mathematical models and in vivo.* PLoS One, 12*(6), e0179131. https://doi.org/10.1371/journal.pone.0179131 under CC BY 4.0 license.*

and (2) that the locally delivered dose can be modulated based on heating duration. That means that TSLs provide the ability to control both the location of delivery and the dose delivered to the targeted tissue.

4. ENCAPSULATED DRUGS AND TARGETED TUMOR TYPES

A wide variety of drugs have been encapsulated in TSL combinations and have been tested in different in vitro and in vivo models. Even though the initial compound encapsulated by (Yatvin et al. (1978) was not for the treatment of cancers, chemotherapeutic drugs were encapsulated almost immediately thereafter (Weinstein, Magin, Yatvin, & Zaharko, 1979). Several different classes of drugs have been encapsulated since then. These include alkylating agents (melphalan (Chelvi & Ralhan, 1997), cisplatin (Dohmitsu et al., 1991; Liu et al., 2015; Yatvin, Muhlensiepen, Porschen, Weinstein, & Feinendegen, 1981), oxaliplatin (Zeng et al., 2016)), plant alkaloids (vinorelbine (Wang, Mei, et al., 2016; Zhang et al., 2011), vincristine (Li et al., 2016), paclitaxel (Wang, Zhang, et al., 2016), docetaxel (Zhang et al., 2014)), antitumor antibiotics (several anthracyclines (Gaber et al., 1995; Lu, Lokerse, Seynhaeve, Koning, & ten Hagen, 2015; Needham et al., 2000; Tomita et al., 1989; van Bree, Krooshoop, Rietbroek, Kipp, & Bakker, 1996; Wu et al., 2011) and bleomycin), and antimetabolites (methotrexate (Weinstein et al., 1979), 5-fluorouracil (Al Sabbagh et al., 2015), gemcitabine (Limmer et al., 2014). The various cancer therapeutics that have been encapsulated by TSLs are listed in Table 1. More detailed information on the history and development of TSLs is discussed elsewhere (Motamarry, Asemani, & Haemmerich, 2017).

So far only a single TSL formulation (encapsulating doxorubicin) has made it to human clinical trials, for the treatment of liver cancer (phase I/III) (Lyon et al., 2017; Tak et al., 2017), for recurrent chest wall breast cancer (phase I) (Zagar et al., 2014), and for pediatric solid tumors (phase I) (Kim et al., 2017). The ability to encapsulate different agents presents therapeutic potential for a variety of tumor types, and this has been studied in vitro and in vivo:

a. *Liver cancer*: Needham et al.'s (2000) formulation of TSL-Dox has been tested in the clinic in combination with radiofrequency ablation (RFA) since the 2000s (Poon & Borys, 2009, 2011; Wood et al., 2012). This formulation failed in a first phase III clinical trial (Tak et al., 2017). Since a post hoc analysis of this trial showed superior treatment outcome when

Table 1 Various TSL Formulations Have Been Employed to Encapsulate a Number of Different Anticancer Agents

Lipid Composition	Lipid Ratios	Loaded Drug	Drug Class/Subclass	Reference
Phosphatidyl choline:Cho	Not available	Melphalan	Alkylating Agent/Mustard gas derivative	Chelvi and Ralhan (1997)
DPPC:DSPC	7:1	Cisplatin	Alkylating Agent/Metal Salt	Dohmitsu et al. (1991), Liu et al. (2015), and Yatvin et al. (1981)
DPPC:MSPC:poloxamer 188: DSPE-PEG2000	85:9.5:0.5:5	Oxaliplatin	Alkylating Agent/Metal Salt	Zeng et al. (2016)
DPPC: MPPC: DSPE-PEG2000	86:5:4	Vinorelbine	Alkaloids/Vinca	Wang, Mei, et al. (2016) and Zhang et al. (2011)
DPPC:DSPE-PEG2000:MSPC	75:17:8	Vincristine	Alkaloids/Vinca	Li et al. (2016)
DPPC:MSPC:DSPE-PEG2000: DSPG	83:3:10:4	Paclitaxel	Alkaloids/Taxanes	Wang, Zhang, et al. (2016)
DPPC:DSPE-PEG2000:EPC: MSPC:DTX	82:11:4:3:4	Docetaxel	Alkaloids/Taxanes	Zhang et al. (2014)
DPPC:Cho	Not available	Doxorubicin	Antitumor antibiotics/Anthracyclines	Tomita et al. (1989)
DPPC:HSPC:Cho:PEG	100:50:30:6	Doxorubicin	Antitumor antibiotics/Anthracyclines	Gaber et al. (1995)
DSPC:Cho	Not available	Daunorubicin	Antitumor antibiotics/Anthracyclines	van Bree et al. (1996)
DPPC:DSPC:DSPE-PEG	6:3.5:0.5	Idarubicin	Antitumor antibiotics/Anthracyclines	Lu et al. (2015)
DPPC:MSPC:DSPG:DSPE-mPEG2000	82:8:10:4	Epirubicin	Antitumor antibiotics/Anthracyclines	Wu et al. (2011)
DPPC:DSPC	9:1	Bleomycin	Antitumor antibiotics/Miscellaneous	Tiwari, Udupa, Rao, and Devi (2000)
DPPC:DSPC	7:3	Methotrexate	Antimetabolites/Folic acid antagonist	Weinstein et al. (1979)
DPPC:CHO:DSPE-PEG	90:5:5	5-Fluorouracil	Antimetabolites/Pyrimidine Antagonist	Al Sabbagh et al. (2015)
DPPC:DSPC:DPPG2	50:20:30	Gemcitabine	Antimetabolites/Pyrimidine antagonist	Limmer et al., (2014)
DPPC:HSPC:Cho	55.6:27.8:16.6	Sorafenib	Kinase Inhibitor/Tyrosine kinase inhibitor	Abshire et al. (2017)

heating duration was longer than 45 min (Lencioni et al., 2014; Tak et al., 2017), with additional follow-up studies in animals confirming benefits of extended heating duration (Rossmann et al., 2017; Swenson et al., 2015), a follow-up phase III trial with optimized heating algorithm has been initiated (Mohamed, Hamad, Hafez, Wooley, & Elsabahy, 2017). Furthermore, a phase I trial for liver cancer patients in combination with high-intensity focused ultrasound hyperthermia has been recently initiated (Lyon et al., 2017).

In addition, Wang et al. showed using thermosensitive liposomal vinorelbine combined with RFA showed improved survival in mice with subcutaneous liver cancer tumors (Wang, Mei, et al., 2016).

b. *Breast cancer*: Needham et al.'s (2000) formulation of TSL-Dox has been tested in clinical phase I/II trials for recurrent chest wall breast cancer (Zagar et al., 2014). Prior in vitro studies demonstrated efficacy of TSL-Dox against drug-resistant breast cancer cells (Merlin, Marchal, Ramacci, Notter, & Vigneron, 1993). Lokerse et al. evaluated TSL-Dox in two orthotopic breast cancer mouse models and showed that efficacy varied depending on cell type (Lokerse et al., 2017). Other drugs that have been tested in breast cancer preclinical models include docetaxel (Zhang et al., 2014) and daunorubicin (Merlin, Marchal, Ramacci, Notter, & Vigneron, 1995). Merlin et al. showed that there is an increased accumulation of daunorubicin within breast cancer cells when used as a TSL formulation along with hyperthermia. Combination studies of TSLs with gold nanoparticles and hyperthermia showed synergy and improved survival in mice carrying breast cancers (Ou et al., 2016).

c. *Lung cancer*: Lewis lung carcinoma is a widely studied cell line for lung cancer evaluation in vitro and in vivo. A variety of different drugs including doxorubicin (Shimose et al., 2001), oxaliplatin (Zeng et al., 2016), paclitaxel (Wang, Zhang, et al., 2016), and cisplatin (Dou et al., 2016) have evaluated and showed an improvement in overall survival of the animals and good treatment response.

d. *Brain cancer*: Chemotherapeutic agents such as doxorubicin (Gong et al., 2011) and cisplatin (Kroin & Penn, 1982) are cytotoxic against brain cancer cells in vitro, but are ineffective in vivo as they cannot cross the blood–brain barrier. It is also known that application of local hyperthermia can enhance the permeability of blood–brain barrier (Kiyatkin & Sharma, 2009), and several studies in small animals demonstrate therapeutic potential of TSL. Kakinuma et al. showed that TSL encapsulated cisplatin can be delivered across the blood–brain barrier, where hyperthermia was used both to open the blood–brain barrier and to release the drug from TSL (Kakinuma, Tanaka, Takahashi, Sekihara, et al., 1996). Additional small animal studies showed efficacy of TSL encapsulated doxorubicin against glioma in vivo (Aoki et al., 2004; Kakinuma, Tanaka, Takahashi, Watanabe, et al., 1996). A recent study in canines demonstrated ability to open the blood-brain barrier byhyperthermia, and deliver doxorubicin across the opened barrier released from TSLs in normal brain (Bredlau et al., 2018).

e. *Head and neck, skin cancer*: Local hyperthermia has been shown to be particularly useful in surface tumors that can be easily accessed such as head and neck cancers (Gao, Zheng, Ren, Tang, & Liang, 2016) and skin cancers such as melanomas (Goyal, Thatai, & Sapra, 2017). Dohmitsu et al. have showed that delivery of TSL-Dox in combination with hyperthermia can be effective against mouse tumor models of head and neck cancers (Dohmitsu et al., 1991). Similarly TSL-idarubicin with mild hyperthermia has been shown to inhibit tumor growth and improve survival in mice models of skin cancer.

f. *Colorectal cancer*: TSL-Dox was shown to be effective against colorectal cancer in vivo in the early 1990s (Maruyama, Unezaki, Takahashi, & Iwatsuru, 1993; Unezaki et al., 1994; Wells, Sen, & Hui, 2003).

g. *Bladder cancer:* Doxorubicin has been in clinical use for bladder cancer. It was recently shown in in vivo porcine studies in normal bladder that TSL-Dox combined with hyperthermia facilitated by warm saline irrigation provided enhanced drug delivery and deeper tissue penetration of drug compared to free drug (Mikhail et al., 2017).

In addition to the studies above encapsulating a single payload, several recent liposome formulations encapsulate more than one compound. Li et al. encapsulated the two chemotherapeutic agents doxorubicin and vincristine (Li et al., 2016). Negussie et al. (2011) and Dou et al. (2015) encapsulated MRI contrast agents along with doxorubicin. This serves the dual purpose of both delivering the drug to the target area by localized hyperthermia and real-time imaging using MRI, discussed in more detail later.

5. IMAGE-GUIDED HYPERTHERMIA

Tumors in human patients are usually identified by medical imaging technologies such as computed tomography (CT), MRI, and ultrasound imaging. In the paradigm of image-guided drug delivery, such medical images are typically used in a first step to identify the tissue region(s) (e.g., tumor(s) or organ(s)) where drug delivery is intended. The use of TSL for image-guided drug delivery requires a heating modality with the ability to expose the targeted region to hyperthermic temperatures in controlled manner—i.e., an image-guided hyperthermia technology is required.

TSLs have been used in combination with several different hyperthermia modalities in animals and human patients. Some of the heating technologies employed were clinically available devices, others were custom devices designed for a specific application. The choice of the hyperthermia device depends primarily on the location and size of the targeted tissue region. Ideally, the targeted tissue will be exposed to temperatures where TSLs release their payload (typically >40°C), with limited heating of nontargeted tissue regions. Temperatures above ~43–45°C can result in reduced perfusion (Rossmann & Haemmerich, 2014) and should ideally be avoided since without blood flow TSL cannot enter the targeted tissues. This means that ideally, the targeted tissue region is heated to the narrow temperature range of ~40–43°C. While this is feasible for small animal tumors, for larger human tumors this often poses a considerable challenge. Below we discuss in more detail some of the hyperthermia technologies that have been employed with focus on those technologies that have been used in human patients.

5.1 Microwave Hyperthermia

Microwave technologies have been widely used to expose large tissue volumes to hyperthermic temperatures (Ryan & Brace, 2017). Depending on location of the target volume, microwaves can be applied from the skin surface via surface applicators or in deep tissue regions via interstitial applicators. The combination of surface microwave applicators with TSL has been the subject of past and on-going clinical trials for treatment of recurrent chest wall cancer (Zagar et al., 2014).

5.2 Radiofrequency Ablation

Tumor ablation has been in clinical use since the 1990s to directly kill tumors by heating them to temperatures above ~50°C. RFA is the most widely employed clinical tumor ablation modality and uses RFA electrodes placed into the tumor guided by medical imaging (e.g., ultrasound or CT). RFA is in clinical use for treatment of cancers in the liver, lung, kidney, and bone, among others (Ahmed, Brace, Lee Jr, & Goldberg, 2011). Local tumor recurrence is problematic particularly when treating larger tumors

(>3 cm in diameter), and these recurrences occur frequently in the margin of the treatment zone. The combination of RFA with TSL may address these margin recurrences by preferentially delivering chemotherapy to the margin region not killed by heat alone (Fig. 7). This combination therapy has been the subject of past and on-going clinical trials (Mohamed et al., 2017; Tak et al., 2017).

5.3 High-Intensity Focused Ultrasound

High-intensity focused ultrasound (HIFU) is a heating technology that has experienced extensive interest within the past decade and is now in clinical use for treatment of uterine fibroids, painful bone metastases, and essential tremor. HIFU employs ultrasound focused in deep tissue regions to facilitate highly localized tissue heating, allowing targeted hyperthermia with mm accuracy. It is often combined with MR thermometry for noninvasive measurement of tissue temperature. MR thermometry uses standard MRI equipment to provide tissue temperature maps in real-time and is thus ideally suited to monitor and control HIFU heating (Rieke & Butts Pauly, 2008). The noninvasiveness, high targeting accuracy, and ability to monitor temperature make HIFU an ideal modality for targeted drug delivery when combined with TSL (Fig. 8). Numerous preclinical studies have investigated therapeutic potential of HIFU and TSL for image-guided drug delivery (Gasselhuber, Dreher, Partanen, et al., 2012; Grull & Langereis, 2012;

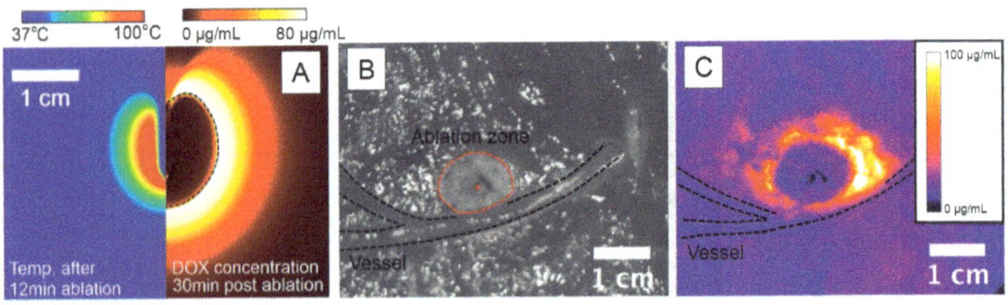

Fig. 7 Combination of tumor ablation with TSL. (A) 2-D computer simulation of temperature (*left*), and of TSL drug delivery (*right*), showing drug uptake in the margin of the ablation zone. Details of the computer model are provided in a prior study (Gasselhuber, Dreher, Negussie, et al., 2010). (B) Results of a recent porcine animal study (normal liver) demonstrating visible ablation zone, and (C) drug delivery in the margin of the ablation zone visualized by fluorescence imaging, qualitatively similar to the computer model prediction in (A). *Reproduced with permission. © 2017, Motamarry A et al. Originally published in Motamarry, A., Asemani, D., & Haemmerich, D. (2017). Chapter 7: Thermosensitive Liposomes. In A. Catala (Ed.),* Liposomes (pp 188–204). *Rijeka: InTech under CC BY 3.0 license.*

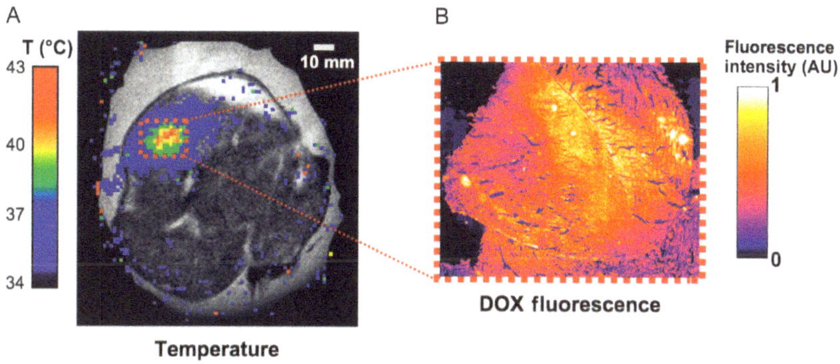

Fig. 8 (A) Temperature map during MRgHIFU-hyperthermia measured via MR thermometry, overlaid on preprocedural proton density-weighted image of rabbit thigh muscle. The targeted tissue region is heated to ~40–42°C after administration of TSL-Dox. (B) Doxorubicin distribution visualized via fluorescence microscopy in extracted tissue sample demonstrating localized, image-guided drug delivery. *Figure adapted from Gasselhuber, A., Dreher, M. R., Partanen, A., Yarmolenko, P. S., Woods, D., Wood, B. J., et al. (2012). Targeted drug delivery by high intensity focused ultrasound mediated hyperthermia combined with temperature-sensitive liposomes: computational modelling and preliminary in vivo validation.* International Journal of Hyperthermia, *28(4), 337–348. https://doi.org/10.3109/02656736.2012.677930.*

Staruch, Hynynen, & Chopra, 2015), and there have been recent clinical trials in adults and pediatric patients (Kim et al., 2017; Lyon et al., 2017).

The combination with MRI has further motivated the coencapsulation of MR contrast agents together with a therapeutic agent (Hijnen et al., 2017; Negussie et al., 2011; Ponce et al., 2007; Ranjan et al., 2012). While it is usually not possible to directly image the drug, the release of the contrast agent provides tissue contrast for monitoring location of delivery, and potentially estimating delivered drug dose (Hijnen et al., 2017; Ponce et al., 2007).

6. SUMMARY

Since the first TSLs have been first been proposed for targeted cancer therapy in the 1970s, they have been widely studied combined with various therapeutic agents, and a first TSL formulation has been evaluated in clinical trials. TSLs offer some unique characteristics that make it attractive for certain cancers that can benefit from locoregional drug delivery: (1) drug concentrations up to ~20–30 times higher compared to therapy with unencapsulated drug may be achievable; (2) with adequate heating modality, high spatial targeting accuracy (~mm) is feasible; (3) TSLs facilitate image-guided drug delivery where targeted tissue region is defined via patient-specific medical images; (4) locally delivered dose can be modulated based on heating duration. Thus, TSL-based image-guided drug delivery has the potential for significant clinical impact in the treatment of cancer.

REFERENCES

Abshire, C., Murad, H. Y., Arora, J. S., Liu, J., Mandava, S. H., John, V. T., et al. (2017). Focused ultrasound-triggered release of tyrosine kinase inhibitor from thermosensitive liposomes for treatment of renal cell carcinoma. *Journal of Pharmaceutical Sciences, 106*(5), 1355–1362. https://doi.org/10.1016/j.xphs.2017.01.027.

Ahmed, M., Brace, C. L., Lee, F. T., Jr., & Goldberg, S. N. (2011). Principles of and advances in percutaneous ablation. *Radiology, 258*(2), 351–369. https://doi.org/10.1148/radiol.10081634.

Al Sabbagh, C., Tsapis, N., Novell, A., Calleja-Gonzalez, P., Escoffre, J. M., Bouakaz, A., et al. (2015). Formulation and pharmacokinetics of thermosensitive stealth(R) liposomes encapsulating 5-fluorouracil. *Pharmaceutical Research, 32*(5), 1585–1603. https://doi.org/10.1007/s11095-014-1559-0.

Aoki, H., Kakinuma, K., Morita, K., Kato, M., Uzuka, T., Igor, G., et al. (2004). Therapeutic efficacy of targeting chemotherapy using local hyperthermia and thermosensitive liposome: Evaluation of drug distribution in a rat glioma model. *International Journal of Hyperthermia, 20*(6), 595–605. https://doi.org/10.1080/02656730410001703186.

Barenholz, Y. (2012). Doxil(R)—The first FDA-approved nano-drug: Lessons learned. *Journal of Controlled Release, 160*(2), 117–134. https://doi.org/10.1016/j.jconrel.2012.03.020.

Bredlau, A. L., Motamarry, A., Chen, C., McCrackin, M. A., Helke, K., Armeson, K. E., et al. (2018). Localized delivery of therapeutic doxorubicin dose across the canine blood–brain barrier with hyperthermia and temperature sensitive liposomes. *Drug Delivery, 25*(1), 973–984. https://doi.org/10.1080/10717544.2018.1461280.

Burke, C., Dreher, M. R., Negussie, A. H., Mikhail, A. S., Yarmolenko, P., Patel, A., et al. (2017). Drug release kinetics of temperature sensitive liposomes measured at high-temporal resolution with a millifluidic device. *International Journal of Hyperthermia*, 1–9. https://doi.org/10.1080/02656736.2017.1412504.

Chelvi, T. P., & Ralhan, R. (1997). Hyperthermia potentiates antitumor effect of thermosensitive-liposome-encapsulated melphalan and radiation in murine melanoma. *Tumour Biology, 18*(4), 250–260.

Deng, Z., Xiao, Y., Pan, M., Li, F., Duan, W., Meng, L., et al. (2016). Hyperthermia-triggered drug delivery from iRGD-modified temperature-sensitive liposomes enhances the anti-tumor efficacy using high intensity focused ultrasound. *Journal of Controlled Release, 243*, 333–341. https://doi.org/10.1016/j.jconrel.2016.10.030.

Dewhirst, M. W., Landon, C. D., Hofmann, C. L., & Stauffer, P. R. (2013). Novel approaches to treatment of hepatocellular carcinoma and hepatic metastases using thermal ablation and thermosensitive liposomes. *Surgical Oncology Clinics of North America, 22*(3), 545–561. https://doi.org/10.1016/j.soc.2013.02.009.

Dicheva, B. M., ten Hagen, T. L., Seynhaeve, A. L., Amin, M., Eggermont, A. M., & Koning, G. A. (2015). Enhanced specificity and drug delivery in tumors by cRGD-anchoring thermosensitive liposomes. *Pharmaceutical Research, 32*(12), 3862–3876. https://doi.org/10.1007/s11095-015-1746-7.

Dohmitsu, K., Akiyama, S., Kawasaki, S., Kataoka, M., Kondoh, K., Yamauchi, M., et al. (1991). Hyper-thermo chemotherapy of esophageal cancer with thermosensitive liposome, TAC-1043. *Gan to Kagaku Ryoho, 18*(11), 1760–1763.

Dou, Y. N., Dunne, M., Huang, H., McKee, T., Chang, M. C., Jaffray, D. A., et al. (2016). Thermosensitive liposomal cisplatin in combination with local hyperthermia results in tumor growth delay and changes in tumor microenvironment in xenograft models of lung carcinoma. *Journal of Drug Targeting, 24*(9), 865–877. https://doi.org/10.1080/1061186x.2016.1191079.

Dou, Y. N., Weersink, R. A., Foltz, W. D., Zheng, J., Chaudary, N., Jaffray, D. A., et al. (2015). Custom-designed laser-based heating apparatus for triggered release of cisplatin from thermosensitive liposomes with magnetic resonance image guidance. *Journal of Visualized Experiments, 106*. e53055. https://doi.org/10.3791/53055.

El-Kareh, A. W., & Secomb, T. W. (2000). A mathematical model for comparison of bolus injection, continuous infusion, and liposomal delivery of doxorubicin to tumor cells. *Neoplasia, 2*(4), 325–338.

Gaber, M. H., Hong, K., Huang, S. K., & Papahadjopoulos, D. (1995). Thermosensitive sterically stabilized liposomes: Formulation and in vitro studies on mechanism of doxorubicin release by bovine serum and human plasma. *Pharmaceutical Research, 12*(10), 1407–1416.

Gao, S., Zheng, M., Ren, X., Tang, Y., & Liang, X. (2016). Local hyperthermia in head and neck cancer: Mechanism, application and advance. *Oncotarget*, 7(35), 57367–57378. https://doi.org/10.18632/oncotarget.10350.

Gasselhuber, A., Dreher, M. R., Negussie, A., Wood, B. J., Rattay, F., & Haemmerich, D. (2010). Mathematical spatio-temporal model of drug delivery from low temperature sensitive liposomes during radiofrequency tumour ablation. *International Journal of Hyperthermia, 26*, 499–513.

Gasselhuber, A., Dreher, M. R., Partanen, A., Yarmolenko, P. S., Woods, D., Wood, B. J., et al. (2012). Targeted drug delivery by high intensity focused ultrasound mediated hyperthermia combined with temperature-sensitive liposomes: Computational modelling and preliminary in vivo validation. *International Journal of Hyperthermia, 28*(4), 337–348. https://doi.org/10.3109/02656736.2012.677930.

Gasselhuber, A., Dreher, M. R., Rattay, F., Wood, B. J., & Haemmerich, D. (2012). Comparison of conventional chemotherapy, stealth liposomes and temperature-sensitive liposomes in a mathematical model. *PLoS One*, 7(10). e47453. https://doi.org/10.1371/journal.pone.0047453.

Gong, W., Wang, Z., Liu, N., Lin, W., Wang, X., Xu, D., et al. (2011). Improving efficiency of adriamycin crossing blood brain barrier by combination of thermosensitive liposomes and hyperthermia. *Biological & Pharmaceutical Bulletin, 34*(7), 1058–1064.

Goyal, N., Thatai, P., & Sapra, B. (2017). Skin cancer: Symptoms, mechanistic pathways and treatment rationale for therapeutic delivery. *Therapeutic Delivery, 8*(5), 265–287. https://doi.org/10.4155/tde-2016-0093.

Grull, H., & Langereis, S. (2012). Hyperthermia-triggered drug delivery from temperature-sensitive liposomes using MRI-guided high intensity focused ultrasound. *Journal of Controlled Release, 161*(2), 317–327. https://doi.org/10.1016/j.jconrel.2012.04.041.

Haeri, A., Zalba, S., Ten Hagen, T. L., Dadashzadeh, S., & Koning, G. A. (2016). EGFR targeted thermosensitive liposomes: A novel multifunctional platform for simultaneous tumor targeted and stimulus responsive drug delivery. *Colloids and Surfaces. B, Biointerfaces, 146*, 657–669. https://doi.org/10.1016/j.colsurfb.2016.06.012.

Hauck, M. L., LaRue, S. M., Petros, W. P., Poulson, J. M., Yu, D., Spasojevic, I., et al. (2006). Phase I trial of doxorubicin-containing low temperature sensitive liposomes in spontaneous canine tumors. *Clinical Cancer Research, 12*(13), 4004–4010. https://doi.org/10.1158/1078-0432.CCR-06-0226.

Hijnen, N., Kneepkens, E., de Smet, M., Langereis, S., Heijman, E., & Grull, H. (2017). Thermal combination therapies for local drug delivery by magnetic resonance-guided high-intensity focused ultrasound. *Proceedings of the National Academy of Sciences of the United States of America*, *114*(24), E4802–E4811. https://doi.org/10.1073/pnas.1700790114.
Hossann, M., Wiggenhorn, M., Schwerdt, A., Wachholz, K., Teichert, N., Eibl, H., et al. (2007). In vitro stability and content release properties of phosphatidylglyceroglycerol containing thermosensitive liposomes. *Biochimica et Biophysica Acta*, *1768*(10), 2491–2499. https://doi.org/10.1016/j.bbamem.2007.05.021.
Hotta, R., Cheng, L., Graham, H. K., Nagy, N., Belkind-Gerson, J., Mattheolabakis, G., et al. (2016). Delivery of enteric neural progenitors with 5-HT4 agonist-loaded nanoparticles and thermosensitive hydrogel enhances cell proliferation and differentiation following transplantation in vivo. *Biomaterials*, *88*, 1–11. https://doi.org/10.1016/j.biomaterials.2016.02.016.
Kakinuma, K., Tanaka, R., Takahashi, H., Sekihara, Y., Watanabe, M., & Kuroki, M. (1996). Drug delivery to the brain using thermosensitive liposome and local hyperthermia. *International Journal of Hyperthermia*, *12*(1), 157–165.
Kakinuma, K., Tanaka, R., Takahashi, H., Watanabe, M., Nakagawa, T., & Kuroki, M. (1996). Targeting chemotherapy for malignant brain tumor using thermosensitive liposome and localized hyperthermia. *Journal of Neurosurgery*, *84*(2), 180–184. https://doi.org/10.3171/jns.1996.84.2.0180.
Kim, A., Sharma, K., Yarmolenko, P., Celik, H., Kaplan, R. N., Dome, J., et al. (2017). Phase 1 trial of lyso-thermosensitive liposomal doxorubicin (LTLD) and magnetic resonance guided high intensity focused ultrasound (MR-HIFU) for pediatric refractory solid tumors. *Journal of Clinical Oncology*, *35*(15 Suppl). TPS10579. https://doi.org/10.1200/JCO.2017.35.15_suppl.TPS10579.
Kiyatkin, E. A., & Sharma, H. S. (2009). Permeability of the blood-brain barrier depends on brain temperature. *Neuroscience*, *161*(3), 926–939. https://doi.org/10.1016/j.neuroscience.2009.04.004.
Kneidl, B., Peller, M., Winter, G., Lindner, L. H., & Hossann, M. (2014). Thermosensitive liposomal drug delivery systems: State of the art review. *International Journal of Nanomedicine*, *9*, 4387–4398. https://doi.org/10.2147/IJN.S49297.
Kong, G., Anyarambhatla, G., Petros, W. P., Braun, R. D., Colvin, O. M., Needham, D., et al. (2000). Efficacy of liposomes and hyperthermia in a human tumor xenograft model: Importance of triggered drug release. *Cancer Research*, *60*(24), 6950–6957.
Kono, K. (2001). Thermosensitive polymer-modified liposomes. *Advanced Drug Delivery Reviews*, *53*(3), 307–319.
Kroin, J. S., & Penn, R. D. (1982). Intracerebral chemotherapy: Chronic microinfusion of cisplatin. *Neurosurgery*, *10*(3), 349–354.
Kullberg, M., Mann, K., & Owens, J. L. (2009). A two-component drug delivery system using Her-2-targeting thermosensitive liposomes. *Journal of Drug Targeting*, *17*(2), 98–107. https://doi.org/10.1080/10611860802471562.
Kunjachan, S., Ehling, J., Storm, G., Kiessling, F., & Lammers, T. (2015). Noninvasive imaging of nanomedicines and nanotheranostics: Principles, progress, and prospects. *Chemical Reviews*, *115*(19), 10907–10937. https://doi.org/10.1021/cr500314d.
Lencioni, R., Tak, W. Y., Chen, M. H., Finn, R. S., Sherman, M., Makris, L., et al. (2014). *Standardized radiofrequency ablation (sRFA) ≥ 45 minutes (m) plus lyso-thermosensitive liposomal doxorubicin (LTLD) for solitary hepatocellular carcinoma (HCC) lesions 3–7 cm: A retrospective analysis of phase III HEAT study*. ASCO.
Li, M., Li, Z., Yang, Y., Wang, Z., Yang, Z., Li, B., et al. (2016). Thermo-sensitive liposome co-loaded of vincristine and doxorubicin based on their similar physicochemical properties had synergism on tumor treatment. *Pharmaceutical Research*, *33*(8), 1881–1898. https://doi.org/10.1007/s11095-016-1924-2.

Li, L., Ten Hagen, T. L., Haeri, A., Soullie, T., Scholten, C., Seynhaeve, A. L., et al. (2013). A novel two-step mild hyperthermia for advanced liposomal chemotherapy. *Journal of Controlled Release*, *174*, 202–208. https://doi.org/10.1016/j.jconrel.2013.11.012.

Limmer, S., Hahn, J., Schmidt, R., Wachholz, K., Zengerle, A., Lechner, K., et al. (2014). Gemcitabine treatment of rat soft tissue sarcoma with phosphatidyldiglycerol-based thermosensitive liposomes. *Pharmaceutical Research*, *31*(9), 2276–2286. https://doi.org/10.1007/s11095-014-1322-6.

Lindner, L. H., Eichhorn, M. E., Eibl, H., Teichert, N., Schmitt-Sody, M., Issels, R. D., et al. (2004). Novel temperature-sensitive liposomes with prolonged circulation time. *Clinical Cancer Research*, *10*(6), 2168–2178.

Liu, H., Zhang, Y., Han, Y., Zhao, S., Wang, L., Zhang, Z., et al. (2015). Characterization and cytotoxicity studies of DPPC:M(2+) novel delivery system for cisplatin thermosensitivity liposome with improving loading efficiency. *Colloids and Surfaces. B, Biointerfaces*, *131*, 12–20. https://doi.org/10.1016/j.colsurfb.2015.04.029.

Lokerse, W. J. M., Bolkestein, M., Dalm, S. U., Eggermont, A. M. M., de Jong, M., Grull, H., et al. (2017). Comparing the therapeutic potential of thermosensitive liposomes and hyperthermia in two distinct subtypes of breast cancer. *Journal of Controlled Release*, *258*, 34–42. https://doi.org/10.1016/j.jconrel.2017.05.005.

Lu, T., Lokerse, W. J., Seynhaeve, A. L., Koning, G. A., & ten Hagen, T. L. (2015). Formulation and optimization of idarubicin thermosensitive liposomes provides ultrafast triggered release at mild hyperthermia and improves tumor response. *Journal of Controlled Release*, *220*(Pt. A), 425–437. https://doi.org/10.1016/j.jconrel.2015.10.056.

Lu, T., & Ten Hagen, T. L. (2017). Inhomogeneous crystal grain formation in DPPC-DSPC based thermosensitive liposomes determines content release kinetics. *Journal of Controlled Release*, *247*, 64–72. https://doi.org/10.1016/j.jconrel.2016.12.030.

Lyon, P. C., Griffiths, L. F., Lee, J., Chung, D., Carlisle, R., Wu, F., et al. (2017). Clinical trial protocol for TARDOX: A phase I study to investigate the feasibility of targeted release of lyso-thermosensitive liposomal doxorubicin (ThermoDox®) using focused ultrasound in patients with liver tumours. *Journal of Therapeutic Ultrasound*, *5*(1), 28. https://doi.org/10.1186/s40349-017-0104-0.

Magin, R. L., Hunter, J. M., Niesman, M. R., & Bark, G. A. (1986). Effect of vesicle size on the clearance, distribution, and tumor uptake of temperature-sensitive liposomes. *Cancer Drug Delivery*, *3*(4), 223–237.

Maruyama, K., Unezaki, S., Takahashi, N., & Iwatsuru, M. (1993). Enhanced delivery of doxorubicin to tumor by long-circulating thermosensitive liposomes and local hyperthermia. *Biochimica et Biophysica Acta*, *1149*(2), 209–216.

Merlin, J. L., Marchal, S., Ramacci, C., Notter, D., & Vigneron, C. (1993). Antiproliferative activity of thermosensitive liposome-encapsulated doxorubicin combined with 43 degrees C hyperthermia in sensitive and multidrug-resistant MCF-7 cells. *European Journal of Cancer*, *29a*(16), 2264–2268.

Merlin, J. L., Marchal, S., Ramacci, C., Notter, D., & Vigneron, C. (1995). Modulation of daunorubicin intracellular accumulation in P-glycoprotein expressing MCF-7 human breast adenocarcinoma cells by thermosensitive-liposome encapsulation and hyperthermia. *International Journal of Hyperthermia*, *11*(6), 855–865. https://doi.org/10.3109/02656739509052341.

Mikhail, A. S., Negussie, A. H., Pritchard, W. F., Haemmerich, D., Woods, D., Bakhutashvili, I., et al. (2017). Lyso-thermosensitive liposomal doxorubicin for treatment of bladder cancer. *International Journal of Hyperthermia*, *33*(7), 733–740. https://doi.org/10.1080/02656736.2017.1315459.

Mohamed, N. K., Hamad, M. A., Hafez, M. Z., Wooley, K. L., & Elsabahy, M. (2017). Nanomedicine in management of hepatocellular carcinoma: Challenges and opportunities. *International Journal of Cancer*, *140*(7), 1475–1484. https://doi.org/10.1002/ijc.30517.

Motamarry, A., Asemani, D., & Haemmerich, D. (2017). Chapter 7: Thermosensitive liposomes. In A. Catala (Ed.), *Liposomes*. Rijeka: InTech. (pp. 188–204).
Mumtaz, S., Ghosh, P. C., & Bachhawat, B. K. (1991). Design of liposomes for circumventing the reticuloendothelial cells. *Glycobiology*, *1*(5), 505–510.
Na, K., Lee, S. A., Jung, S. H., Hyun, J., & Shin, B. C. (2012). Elastin-like polypeptide modified liposomes for enhancing cellular uptake into tumor cells. *Colloids and Surfaces. B, Biointerfaces*, *91*, 130–136. https://doi.org/10.1016/j.colsurfb.2011.10.051.
Needham, D., Anyarambhatla, G., Kong, G., & Dewhirst, M. W. (2000). A new temperature-sensitive liposome for use with mild hyperthermia: Characterization and testing in a human tumor xenograft model. *Cancer Research*, *60*(5), 1197–1201.
Negussie, A. H., Yarmolenko, P. S., Partanen, A., Ranjan, A., Jacobs, G., Woods, D., et al. (2011). Formulation and characterisation of magnetic resonance imageable thermally sensitive liposomes for use with magnetic resonance-guided high intensity focused ultrasound. *International Journal of Hyperthermia*, *27*(2), 140–155. https://doi.org/10.3109/02656736.2010.528140.
Ou, Y. C., Webb, J. A., Faley, S., Shae, D., Talbert, E. M., Lin, S., et al. (2016). Gold nanoantenna-mediated photothermal drug delivery from thermosensitive liposomes in breast cancer. *ACS Omega*, *1*(2), 234–243. https://doi.org/10.1021/acsomega.6b00079.
Papahadjopoulos, D., Allen, T. M., Gabizon, A., Mayhew, E., Matthay, K., Huang, S. K., et al. (1991). Sterically stabilized liposomes: Improvements in pharmacokinetics and antitumor therapeutic efficacy. *Proceedings of the National Academy of Sciences of the United States of America*, *88*(24), 11460–11464.
Park, K. (2013). Facing the truth about nanotechnology in drug delivery. *ACS Nano*, 7(9), 7442–7447. https://doi.org/10.1021/nn404501g.
Ponce, A. M., Viglianti, B. L., Yu, D., Yarmolenko, P. S., Michelich, C. R., Woo, J., et al. (2007). Magnetic resonance imaging of temperature-sensitive liposome release: Drug dose painting and antitumor effects. *Journal of the National Cancer Institute*, *99*(1), 53–63. https://doi.org/10.1093/jnci/djk005.
Poon, R. T., & Borys, N. (2009). Lyso-thermosensitive liposomal doxorubicin: A novel approach to enhance efficacy of thermal ablation of liver cancer. *Expert Opinion on Pharmacotherapy*, *10*(2), 333–343. https://doi.org/10.1517/14656560802677874.
Poon, R. T., & Borys, N. (2011). Lyso-thermosensitive liposomal doxorubicin: An adjuvant to increase the cure rate of radiofrequency ablation in liver cancer. *Future Oncology*, 7(8), 937–945. https://doi.org/10.2217/fon.11.73.
Ranjan, A., Jacobs, G. C., Woods, D. L., Negussie, A. H., Partanen, A., Yarmolenko, P. S., et al. (2012). Image-guided drug delivery with magnetic resonance guided high intensity focused ultrasound and temperature sensitive liposomes in a rabbit Vx2 tumor model. *Journal of Controlled Release*, *158*(3), 487–494.
Rieke, V., & Butts Pauly, K. (2008). MR thermometry. *Journal of Magnetic Resonance Imaging*, *27*(2), 376–390. https://doi.org/10.1002/jmri.21265.
Rossmann, C., & Haemmerich, D. (2014). Review of temperature dependence of thermal properties, dielectric properties, and perfusion of biological tissues at hyperthermic and ablation temperatures. *Critical Reviews in Biomedical Engineering*, *42*(6), 467–492.
Rossmann, C., McCrackin, M. A., Armeson, K. E., & Haemmerich, D. (2017). Temperature sensitive liposomes combined with thermal ablation: Effects of duration and timing of heating in mathematical models and in vivo. *PLoS One*, *12*(6), e0179131. https://doi.org/10.1371/journal.pone.0179131.
Ryan, T. P., & Brace, C. L. (2017). Interstitial microwave treatment for cancer: Historical basis and current techniques in antenna design and performance. *International Journal of Hyperthermia*, *33*(1), 3–14. https://doi.org/10.1080/02656736.2016.1214884.

Sadeghi, N., Deckers, R., Ozbakir, B., Akthar, S., Kok, R. J., Lammers, T., et al. (2017). Influence of cholesterol inclusion on the doxorubicin release characteristics of lysolipid-based thermosensitive liposomes. *International Journal of Pharmaceutics*. https://doi.org/10.1016/j.ijpharm.2017.11.002.

Shimose, S., Sugita, T., Nitta, Y., Kubo, T., Ikuta, Y., & Murakami, T. (2001). Effect of thermosensitive liposomal doxorubicin with hyperthermia on primary tumor and lung metastases in hamster osteosarcoma. *International Journal of Oncology*, *19*(3), 585–589.

Staruch, R. M., Hynynen, K., & Chopra, R. (2015). Hyperthermia-mediated doxorubicin release from thermosensitive liposomes using MR-HIFU: Therapeutic effect in rabbit Vx2 tumours. *International Journal of Hyperthermia*, *31*(2), 118–133. https://doi.org/10.3109/02656736.2014.992483.

Sullivan, S. M., & Huang, L. (1985). Preparation and characterization of heat-sensitive immunoliposomes. *Biochimica et Biophysica Acta*, *812*(1), 116–126.

Swenson, C. E., Haemmerich, D., Maul, D. H., Knox, B., Ehrhart, N., & Reed, R. A. (2015). Increased duration of heating boosts local drug deposition during radiofrequency ablation in combination with thermally sensitive liposomes (ThermoDox) in a porcine model. *PLoS One*, *10*(10), e0139752. https://doi.org/10.1371/journal.pone.0139752.

Tak, W. Y., Lin, S. M., Wang, Y., Zheng, J., Vecchione, A., Park, S. Y., et al. (2017). Phase III HEAT study adding lyso-thermosensitive liposomal doxorubicin to radiofrequency ablation in patients with unresectable hepatocellular carcinoma lesions. *Clinical Cancer Research*, *24*, 73–83. https://doi.org/10.1158/1078-0432.CCR-16-2433.

Tiwari, S. B., Udupa, V. N., Rao, S., & Devi, P. U. (2000). Thermochemotherapy: Synergism between hyperthermia and liposomal bleomycin in mice bearing melanoma B16F1. *Pharmacy and Pharmacology Communications*, *6*(1), 19–23. https://doi.org/10.1211/146080800128735421.

Tomita, T., Watanabe, M., Takahashi, T., Kumai, K., Tadakuma, T., & Yasuda, T. (1989). Temperature-sensitive release of adriamycin, an amphiphilic antitumor agent, from dipalmitoylphosphatidylcholine-cholesterol liposomes. *Biochimica et Biophysica Acta*, *978*(2), 185–190.

Unezaki, S., Maruyama, K., Takahashi, N., Koyama, M., Yuda, T., Suginaka, A., et al. (1994). Enhanced delivery and antitumor activity of doxorubicin using long-circulating thermosensitive liposomes containing amphipathic polyethylene glycol in combination with local hyperthermia. *Pharmaceutical Research*, *11*(8), 1180–1185.

van Bree, C., Krooshoop, J. J., Rietbroek, R. C., Kipp, J. B., & Bakker, P. J. (1996). Hyperthermia enhances tumor uptake and antitumor efficacy of thermostable liposomal daunorubicin in a rat solid tumor. *Cancer Research*, *56*(3), 563–568.

Vingerhoeds, M. H., Steerenberg, P. A., Hendriks, J. J., Dekker, L. C., Van Hoesel, Q. G., Crommelin, D. J., et al. (1996). Immunoliposome-mediated targeting of doxorubicin to human ovarian carcinoma in vitro and in vivo. *British Journal of Cancer*, *74*(7), 1023–1029.

Wang, S., Mei, X. G., Goldberg, S. N., Ahmed, M., Lee, J. C., Gong, W., et al. (2016). Does thermosensitive liposomal vinorelbine improve end-point survival after percutaneous radiofrequency ablation of liver tumors in a mouse model? *Radiology*, *279*(3), 762–772. https://doi.org/10.1148/radiol.2015150787.

Wang, C., Wang, X., Zhong, T., Zhao, Y., Zhang, W. Q., Ren, W., et al. (2015). The antitumor activity of tumor-homing peptide-modified thermosensitive liposomes containing doxorubicin on MCF-7/ADR: In vitro and in vivo. *International Journal of Nanomedicine*, *10*, 2229–2248. https://doi.org/10.2147/ijn.s79840.

Wang, Z. Y., Zhang, H., Yang, Y., Xie, X. Y., Yang, Y. F., Li, Z., et al. (2016). Preparation, characterization, and efficacy of thermosensitive liposomes containing paclitaxel. *Drug Delivery*, *23*(4), 1222–1231. https://doi.org/10.3109/10717544.2015.1122674.

Weinstein, J. N., Magin, R. L., Yatvin, M. B., & Zaharko, D. S. (1979). Liposomes and local hyperthermia: Selective delivery of methotrexate to heated tumors. *Science*, *204*(4389), 188–191.

Wells, J., Sen, A., & Hui, S. W. (2003). Localized delivery to CT-26 tumors in mice using thermosensitive liposomes. *International Journal of Pharmaceutics*, *261*(1–2), 105–114.

Wood, B. J., Poon, R. T., Locklin, J. K., Dreher, M. R., Ng, K. K., Eugeni, M., et al. (2012). Phase I study of heat-deployed liposomal doxorubicin during radiofrequency ablation for hepatic malignancies. *Journal of Vascular and Interventional Radiology*, *23*(2), 248–255, e247. https://doi.org/10.1016/j.jvir.2011.10.018.

Wu, Y., Yang, Y., Zhang, F. C., Wu, C., Lu, W. L., & Mei, X. G. (2011). Epirubicin-encapsulated long-circulating thermosensitive liposome improves pharmacokinetics and antitumor therapeutic efficacy in animals. *Journal of Liposome Research*, *21*(3), 221–228. https://doi.org/10.3109/08982104.2010.520273.

Yatvin, M. B., Muhlensiepen, H., Porschen, W., Weinstein, J. N., & Feinendegen, L. E. (1981). Selective delivery of liposome-associated cis-dichlorodiammineplatinum(II) by heat and its influence on tumor drug uptake and growth. *Cancer Research*, *41*(5), 1602–1607.

Yatvin, M. B., Weinstein, J. N., Dennis, W. H., & Blumenthal, R. (1978). Design of liposomes for enhanced local release of drugs by hyperthermia. *Science*, *202*(4374), 1290–1293.

Zagar, T. M., Vujaskovic, Z., Formenti, S., Rugo, H., Muggia, F., O'Connor, B., et al. (2014). Two phase I dose-escalation/pharmacokinetics studies of low temperature liposomal doxorubicin (LTLD) and mild local hyperthermia in heavily pretreated patients with local regionally recurrent breast cancer. *International Journal of Hyperthermia*, *30*(5), 285–294. https://doi.org/10.3109/02656736.2014.936049.

Zeng, C., Yu, F., Yang, Y., Cheng, X., Liu, Y., Zhang, H., et al. (2016). Preparation and evaluation of oxaliplatin thermosensitive liposomes with rapid release and high stability. *PLoS One*, *11*(7), e0158517. https://doi.org/10.1371/journal.pone.0158517.

Zhang, H., Gong, W., Wang, Z. Y., Yuan, S. J., Xie, X. Y., Yang, Y. F., et al. (2014). Preparation, characterization, and pharmacodynamics of thermosensitive liposomes containing docetaxel. *Journal of Pharmaceutical Sciences*, *103*(7), 2177–2183. https://doi.org/10.1002/jps.24019.

Zhang, K., Liu, M., Tong, X., Sun, N., Zhou, L., Cao, Y., et al. (2015). Aptamer-modified temperature-sensitive liposomal contrast agent for magnetic resonance imaging. *Biomacromolecules*, *16*(9), 2618–2623. https://doi.org/10.1021/acs.biomac.5b00250.

Zhang, H., Wang, Z. Y., Gong, W., Li, Z. P., Mei, X. G., & Lv, W. L. (2011). Development and characteristics of temperature-sensitive liposomes for vinorelbine bitartrate. *International Journal of Pharmaceutics*, *414*(1–2), 56–62. https://doi.org/10.1016/j.ijpharm.2011.05.013.

Zhao, F., Zhou, J., Su, X., Wang, Y., Yan, X., Jia, S., et al. (2017). A smart responsive dual aptamers-targeted bubble-generating nanosystem for cancer triplex therapy and ultrasound imaging. *Small*, *13*(20). https://doi.org/10.1002/smll.201603990.

Zimmermann, K., Hossann, M., Hirschberger, J., Troedson, K., Peller, M., Schneider, M., et al. (2017). A pilot trial of doxorubicin containing phosphatidyldiglycerol based thermosensitive liposomes in spontaneous feline soft tissue sarcoma. *International Journal of Hyperthermia*, *33*, 128–190. https://doi.org/10.1080/02656736.2016.1230233.

CHAPTER SIX

Supramolecular Analytical Chemistry in Cancer Research

Shiqun Shao, Min Xue[1]
Department of Chemistry, University of California Riverside, Riverside, CA, United States
[1]Corresponding author: e-mail address: min.xue@ucr.edu

Contents

Abstract

Supramolecular interactions, such as those observed between antibodies and antigens, have been employed in developing analytical methods for several decades. One major area of interest concerns cancer research, where intricate supramolecular designs have emerged to tackle difficult analytes in complex tumor systems. Our increasing knowledge toward supramolecular systems have elicited profound interest in creating more efficient analytical approaches, evidenced by the ever-growing body of literature in the field. Some of the novel tools have indeed facilitated our understanding of cancer biology, through providing previously inaccessible information. In this review, we describe common strategies of developing supramolecular analytical methods and their implementations in cancer research. We provide an overview for each of the approaches and discuss representative examples in recent literature.

1. INTRODUCTION

Since Hermann Fischer's initial "lock and key" theory on enzymes, understanding supramolecular interactions has become pivotal in studying biological systems (Cragg, 2010). Noncovalent interactions among molecular entities (hence the term supramolecular), such as those involving hydrogen bonding and van der Waals force, dominate almost all aspects

Advances in Cancer Research, Volume 139
ISSN 0065-230X
https://doi.org/10.1016/bs.acr.2018.04.005

of biology. Moving along the central dogma of molecular biology, the prevailing contribution from supramolecular processes is rather obvious. The storage of genetic information in DNA relies on the hydrogen bonds between complementary base pairs and the stacking of aromatic rings. Accurate transcription and translation of the genetic information require intricate protein machinery, whose precise assembly is completely governed by supramolecular recognition. The panoply of protein–protein and protein–metabolite interactions that maintain normal cellular function and homeostasis also follow basic supramolecular principles. Indeed, nature has embedded and perfected the art of supramolecular control throughout the evolution of life.

Accompanying the expansion of our knowledge on natural supramolecular interactions, our attempts in creating complex artificial supramolecular assemblies have gained great progress (Wei, Yan, & Huang, 2015). From the synthesis of crown ethers to the seemingly impossible realization of mechanically interlocked molecules, these endeavors have intrigued numerous scientists (Cheng et al., 2015; van Dongen, Cantekin, Elemans, Rowan, & Nolte, 2014). One ever-growing field concerns the design and implementation of novel supramolecular tools for studying and ultimately manipulating biological processes (Ma & Zhao, 2015). Examples such as cyclodextrin-based drug delivery systems have marked prominent value and success in this direction (Davis, 2009; Ribas et al., 2010; Xue & Zink, 2013; Zhang & Ma, 2013). While these drug delivery systems have paved the way for clinics, efforts toward exploring other biological applications of supramolecular chemistry have also flourished. One prominent field is supramolecular bioanalytical chemistry (Anslyn, 2007; Chinen et al., 2015; Ren, Zhang, Chen, & Yang, 2014; You, Zha, & Anslyn, 2015), where novel methods developed with supramolecular principles enable detailed analysis of biological samples. A large portion of the supramolecular analytical methods is targeting cancer research, where many studies have demonstrated advantages of supramolecular approaches over conventional analytical techniques. Most importantly, while traditional analytical methods are often destructive to analytes or biological process, supramolecular approaches could be nondestructive, making it more suitable for bioanalysis in living cells and in vivo. In addition, due to the versatility of supramolecular chemistry, simultaneous quantification of multiple analytes of different categories can be achieved using orthogonal schemes. Such multiplexity used to be limited to antibody-based immunomethods, and its implementation through supramolecular approaches has greatly expanded the arsenal of bioanalytical techniques.

In this review, we provide an overview of the development of supramolecular analytical chemistry for cancer detection and analysis. Many of the early achievements in this field have been thoroughly reviewed by others (You et al., 2015), therefore, we will focus on recent advances. We categorize the studies based on their mechanisms of action and corresponding supramolecular principles. These categories include: (1) self-assembly/disassembly based detection, (2) conformation change induced signal, (3) competitive binding enabled readout, and (4) direct supramolecular recognition of analytes. For each category, we provide a brief discussion of its unique features and typical examples demonstrating such properties. We do realize that this compartmentalization method is rather coarse and that many of the studies may span across several categories. However, it is not our ultimate goal to provide an unambiguous classification for each study. Instead, we aim to focus on accounting the progress concerning various techniques.

2. DETECTION SCHEMES BASED ON SELF-ASSEMBLY/DISASSEMBLY

The self-assembly process is probably the most studied area of supramolecular chemistry. The formation of nanostructures such as micelles and vesicles has been explored in numerous studies (Zhu, Yang, Chen, & Du, 2017). The basic thermodynamic mechanisms of this process have been well elucidated and implemented in the designing of successful supramolecular drug delivery systems (Davis, Chen, & Shin, 2008; Kataoka, Harada, & Nagasaki, 2012; Peer et al., 2007). Because of the assembly process, the monomer and the corresponding supramolecular structure usually exhibit drastically different physical and chemical properties. This feature can be employed for designing analytical methods, where various external stimuli readily control the self-assembly and disassembly process. Fig. 1 illustrates the mechanisms of action in representative approaches.

Many organic fluorophores undergo severe fluorescence quenching in an aggregated state. This property can, therefore, be easily adapted to create fluorescence turn-on type detection methods, through an analyte-induced disassociation process. For instance, Tan et al. developed a series of near-IR γ-phenyl-substituted Cy5 fluorophores, which can self-assemble into fluorescence-quenched J-type aggregates in the absence of target analytes.

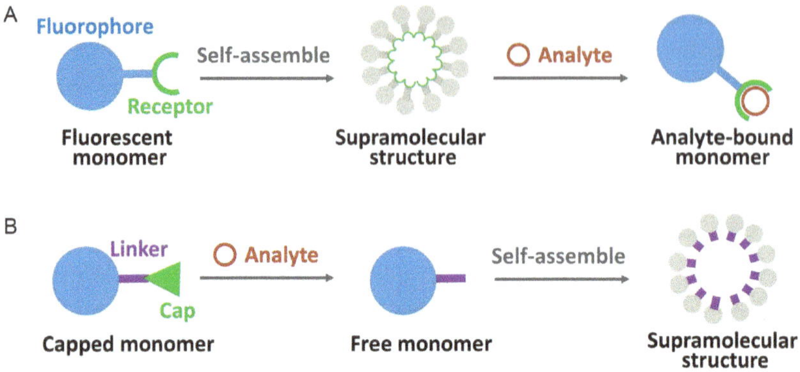

Fig. 1 Schematic illustration of self-disassembly and disassembly based detection principles. (A) Fluorescent monomers consisting of fluorophore and receptors can self-assemble into a supramolecular structure, which exhibits different fluorescence properties. Analyte binding can disrupt this supramolecular assembly, resulting in a change in the fluorescence signal. (B) Analytes can activate capped monomers by chemical modifications, which enables self-assembly of the free monomers. The formation of a supramolecular structure often accompanies changes in optical properties, which indirectly provides readouts from the analytes.

In the presence of the specific analyte, the recognition-induced disassembly of the probe aggregate would take place and greatly enhance the emission of the Cy5 monomer. In particular, one of the probes, benzylamine sulfonamide-conjugated γ-phenyl-substituted Cy5 can be used for wash-free detection of intracellular transmembrane-type carbonic anhydrase IX, a hypoxia-induced cancer-specific biomarker (Hou, Wu, Chiang, & Tan, 2015). Similar realization of disassembly caused fluorescence turn-on assays have also been developed to detect glutathione (Liu, Niu, Chen, & Yang, 2018), α-fetoprotein (He et al., 2015), heat shock protein 90 (Yoshii, Mizusawa, Takaoka, & Hamachi, 2014), and proteases activities (Lin, Chapman, & Stevens, 2014; Yuan, Zhang, Cao, An, & Liang, 2015).

Contrary to conventional fluorophores such as Cy5, luminogens with aggregation-induced emission characteristics (AIEgen) are nonfluorescent as molecular species but highly fluorescent in aggregated states (Mei, Leung, Kwok, Lam, & Tang, 2015). This property renders AIEgens ideal for the detection of cancer-related biomarkers, where the interactions between the probe and biomolecules readily cause aggregation (Kwok, Leung, Lam, & Tang, 2015). Ding and coworkers developed a general strategy to design AIE-based probes by introducing a sequence of self-assembling peptide GFFY between the recognition moiety and the AIEgen (Han et al., 2016). In the presence of the target analytes, the sequence GFFY would be exposed and induce the ordered self-assembly of AIEgens, activating the fluorescence emission. By employing tetraphenylethene (TPE) as the AIEgen, they prepared two probes of TPE-GFFYK(DVEDEE-Ac) and TPE-GFFYE-SS-EE, and achieved highly specific and sensitive detection of intracellular caspase-3 and thiol levels, respectively.

Besides the fluorescence "turn-on" strategy, fluorescence "turn-off" can also be utilized for supramolecular bioanalysis. In those studies, analytes cause the aggregation of fluorophores, which leads to the loss of fluorescence intensity. This approach follows the same principle as those "turn-on" type assays, however, the induction of the aggregated phase as the end point offers different opportunities. For example, Liang and colleagues reported on a sol–gel phase transition-based fluorescence "turn-off" method for the quantification of intracellular alkaline phosphatase activities. A fluorescein-labeled phosphor–peptide Fmoc-K(FITC)-FFYp was prepared as a substrate to alkaline phosphatase. Enzymatic cleavage of the phosphate group leads to the formation of a hydrogel and quenched FITC fluorescence. This method can quantify alkaline phosphatase activity over the range of 0–2.8 U/mL with a detection limit of 0.06 U/mL. Such sensitivity paves the way for early detection of cancer biomarkers (Dong, Miao, Hai, Yuan, & Liang, 2015).

In addition to small molecules, nanoparticles can also serve as building blocks for supramolecular structures. Such design strategies often harness the optical properties of nanoparticles to generate readouts. Examples include the plasmonic effect of gold nanoparticles and the energy harvest properties of upconversion nanoparticles. Kuang and coworkers constructed DNA-driven gold-upconversion nanoparticle (Au-UCNP) pyramids, which display strong plasmonic circular dichroism (CD) at 521 nm and significant luminescence in 500–600 nm. The presence of miRNA disassembles the pyramid nanostructures, resulting in the decrease of CD and increase of luminescence intensity. The experimental results showed that both CD and luminescence intensity linearly correlate with miRNA amounts. Furthermore, the CD detection mode exhibited an impressive detection limit of 0.03 fmol/10 μg_{RNA} (Li et al., 2016). Despite the successful demonstration of supramolecular analytical methods involving nanoparticles, the implementation is often remarkably challenging. The nanoparticle uniformity is crucial to most studies, and routinely synthesizing highly uniform nanoparticles has always been a nontrivial task. Nanoparticles also have strong tendency to aggregate, which will inevitably alter their optical properties. As a result, very few studies have been reported. Nevertheless, nanoparticles provide highly versatile platforms for analyte detection and the prospect is attractive.

3. CONFORMATION CHANGE INDUCED SIGNAL

Many biological macromolecules adopt different conformations upon binding to other molecules or undergoing chemical modifications. Prominent examples include the allosteric control of protein structure and target-assisted rehybridization of DNA aptamers (Alhadrami, Chinnappan, Eissa, Rahamn, & Zourob, 2017; Hu et al., 2015; Lin et al., 2015; Panda et al., 2015; Pittolo et al., 2014; Wang, Mao, et al., 2016; Zhao, Wu, Zhang, & Cai, 2015). Such conformation changes often result in the formation or destruction of molecular moieties with unique features. Strategically targeting these moieties enables detection methods that are both specific and sensitive. Fig. 2 summarizes representative approaches based on conformation change principles.

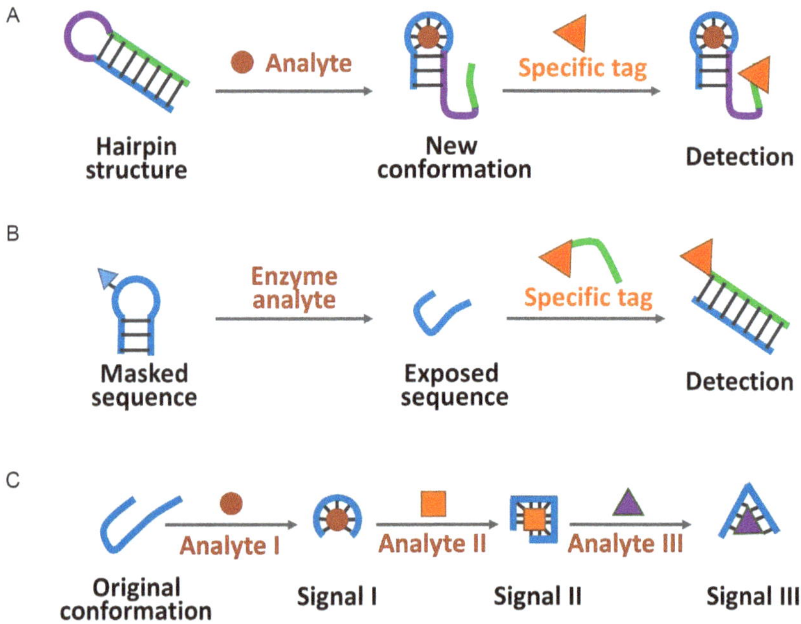

Fig. 2 Schematic illustration of three representative approaches based on conformation changes. (A) DNA or RNA aptamers can adopt a semistable hairpin structure. Analyte binding induces a conformation change and exposes part of the sequence, allowing a specific tag to recognize and provide signal readouts. (B) Chemical modifications on the oligonucleotides can be removed by enzymatic processes. The new sequence can be recognized by a specific tag. (C) Some molecular entities can bind to various analytes while adopting different conformations. Each of the conformations exhibits unique optical properties and therefore enables dynamic quantitation of the analytes.

Owing to the high binding affinity and specificity toward their target molecules, DNA aptamers have been widely used for the construction of analytical detection platforms. Most of the early applications focused on developing monocomponent antibody surrogates, where aptamers specifically recognized analytes and enabled analytical schemes similar to those immuno-based approaches. In recent years, new strategies featuring more complex and versatile mechanisms have emerged. Multicomponent aptamers with semistable conformations can form new structures when interacting with specific molecules. This process can lead to the exposure of readily recognizable segments and in turn provide analyte-specific readouts. For example, the integration of DNA aptamer and G-quadruplex forming DNA has generated many attractive probes for various analytes (Lin et al., 2015; Panda et al., 2015; Wang, Mao, et al., 2016). In one study, Ma et al. reported on such a probe for the detection of protein tyrosine kinase-7 (PTK7), a tumor suppressor relevant in many types of cancers (Golubkov et al., 2014). In the absence of PTK7, the sgc8 aptamer hybridizes to a G-quadruplex forming sequence and another short complementary sequence to form a hairpin structure with two loops. Upon binding with PTK7, the hairpin structure is destructed and the G-quadruplex forming sequence is released. The liberated G-quadruplex forming sequence then folds into a G-quadruplex structure with the assistance of potassium ions. This structure can be detected by the G-quadruplex selective iridium(III) complex with a strong luminescent response (Lin et al., 2015).

On the other hand, DNA aptamer itself can also be engineered to become desired probes (Alhadrami et al., 2017; Zhao et al., 2015). Zhao et al. established a general strategy to design aptamer probes for highly specific and simultaneous detection of oncoproteins in tumor cells. Their aptamer probes comprise of two parts: a target-specific sequence and an extending spacer at each terminus of the aptamer with complementary bases. The first sequence acts as a recognition unit which binds to the target protein on cell surface, and the extending spacer sequence folds the aptamer in a hairpin structure, enabling the close proximity of a quencher/fluorophore pair, which are labeled at the 5′- and 3′-terminus, respectively. In the absence of the target protein, the probe fluorescence is quenched due to a Förster resonance energy transfer (FRET) process. The target protein induces a conformation change that leads to the separation of the fluorophore from the quencher, which significantly enhances the fluorescence signal. Moreover, through the incorporation of different aptamers and various fluorophore/quencher combinations, they successfully achieved simultaneous detection of multiple tumor-related proteins in living cancer cells (Zhao et al., 2015).

Oligonucleotides can be not only built into aptamers but also used as enzymatic substrates. Specific groups on the oligonucleotide sequence can undergo chemical reactions with the aid from certain enzymes. Chemical modifications to these groups often confer new conformations on the oligonucleotides. As an example, Woon et al. synthesized a methylation-responsive label-free nucleic acid (RNA) probe which changes its conformation in response to the methylation status. When methylated, the RNA probe preferentially forms a hairpin structure. Upon demethylation by an RNA demethylase, a significant shift to a duplex structure will occur, resulting in a profound change in the probe melting temperature. This alteration can be sensitively detected through a differential scanning fluorimetry technique (Yang, Cheong, Mai, Zou, & Woon, 2016). Unlike the conventional antibody-based measurements that mainly focus on the expression level, this approach provides information on RNA demethylase activities. This paradigm change is rather valuable, considering the significance of RNA demethylases in many cancers (Esteller & Pandolfi, 2017; Zhang et al., 2017).

Aside from the DNA/RNA-based macromolecules, some entirely artificial supramolecular constructs also exhibit analyte-specific conformation changes. Unlike the binary on–off scheme of aptamers, those artificial structures can provide multistate or multiplex readouts in response to different stimuli, and can even enable dynamic measurements. A recent example was demonstrated by George et al., where they created an adenosine phosphate (AP)-responsive supramolecular helical structure. This structure is comprised of self-assembled naphthalenediimides with zinc coordinated dipicolylethylenediamines (NDPA) as the phosphate recognition unit. ATP imparts P-helix to the NDPA stacks, and the ribose moieties are oriented at the periphery of the stack. By contrast, ADP and AMP induce the M-helix where the riboses are close to the NDPA cores. This switching of helicity is so fast and significant that this assembly is capable of monitoring the reaction kinetics of enzymatic ATP hydrolysis and the stepwise formation of ADP, AMP, and the final phosphates (Kumar et al., 2014).

4. COMPETITIVE BINDING ENABLED READOUT

The noncovalent nature of interactions confers the reversibility to supramolecular processes. As a result, the equilibrium between the bound and unbound moieties can be perturbed by introducing new components that are structurally similar to the binding partners. This competition on binding sites between various guest molecules provides information on the relative

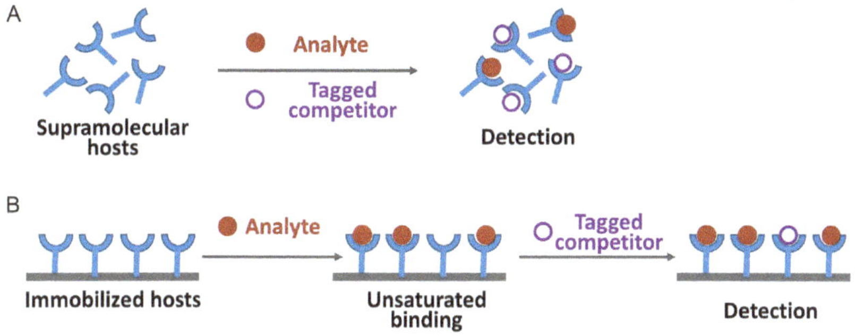

Fig. 3 Schematic illustration of competitive binding-based detection principles. (A) The target analyte competes with a fluorescently tagged surrogate for the binding sites of supramolecular hosts. The competition results in a fluorescence intensity that inversely correlates with the analyte amounts. (B) Surface immobilization of the supramolecular hosts allows sequential recognition.

quantity of each compound, and therefore can be easily harnessed to generate assay readouts. One widely used approach is to fluorescently label the competitor and measure the resulting fluorescence intensity on the host–guest complex, which provides the quantity of the corresponding analyte according to a working curve (Fig. 3). Variations on such strategy have been implemented in numerous studies, and its working principle has been integrated with other supramolecular techniques to enable more complex analytical methods.

Recently, we have developed a panel of surface-based detection schemes targeting metabolites, using competitive binding principles. Intracellular metabolites, such as c-AMP, c-GMP, and glutathione, can be quantified using the competition between fluorescently labeled molecules and native metabolites from cell lysate, toward surface-immobilized antibodies. Energy source molecules, such as glucose and glutamine, can be assayed through a more elaborate approach, implementing supramolecular guest moiety-tagged surrogates. These surrogates can be taken up by the cells and subsequently compete with other optically active guest molecules for the binding sites on the surface-bound supramolecular hosts. This competition enables the quantitation of those surrogates and therefore the cellular metabolic activities (Xue, Wei, Su, Johnson, & Heath, 2016; Xue et al., 2015). By integrating different supramolecular systems, such as antibody–antigen, streptavidin–biotin, and cyclodextrin–adamantane, we were able to perform multiplex quantitation. These approaches can be further miniaturized and adapted to a microfluidic chip platform to enable analysis at single cell level. We have successfully utilized this technology to study the interplay between metabolic and oncogenic protein signaling activities.

The complex of cucurbit[7]uril (CB[7]) and cationic benzylammonium (Benz) is another host–guest system that is frequently exploited in competitive binding assays. Rotello et al. developed a sensing platform which uses the interaction between CB[7] and the Benz-functionalized nanoparticle (BenzNP), to rapidly profile the tumorigenicity of cancer cells. This platform consists of three major parts: BenzNP, three fluorescent proteins (FPs), and CB[7]. In the absence of CB[7], FPs bind with BenzNP due to electrostatic interaction, quenching the fluorescence of FPs. After the addition of cell lysates, different amounts of FPs are released due to their competitive binding for BenzNP, generating three detection channels. The addition of CB[7] significantly changes the fluorescence emission of the three FPs as a result of the competition for the binding sites between the analytes and the newly added CB[7]. This competition generates three additional detection channels. These additional channels allow facile discrimination of cells by their tumorigenicity. Compared to the system without CB[7], BenzNP-CB[7]-FPs system dramatically improved accuracy and efficiency (Le et al., 2017).

Because the competitive binding-based detection principle is rather straightforward, many studies have successfully demonstrated its implementation. For example, Wang et al. implemented a calixpyridinium-based supramolecular assay for the real time, continuous, and label-free monitoring of alkaline phosphatase activity (Wang, Cui, Xing, & Dou, 2016). Sun et al. reported on a sensitive and selective fluorescent method for detection of cholesterol based on cyclodextrin functionalized carbon quantum dots through competitive host–guest recognition (Sun, Fang, Fang, Qian, & Feng, 2017). Krull et al. demonstrated a FRET-based assay on modified cellulose paper using aptamer-linked quantum dots for the detection of epithelial cell adhesion molecules (Das & Krull, 2017). Zhao et al. developed an aptamer-based competitive fluorescence anisotropy/polarization assay for ATP using aptamer as affinity ligand and dye-labeled ATP as fluorescence tracer (Li, Sun, & Zhao, 2017).

5. DIRECT SUPRAMOLECULAR RECOGNITION OF ANALYTES

The most complex yet efficient supramolecular interactions in biological systems are those concerning immunorecognitions. The antibody–antigen binding and the T-cell receptor (TCR)–major histocompatibility complex (MHC) interaction are among the most versatile ones. The ability to specifically bind to a target molecule almost immediately enables its analysis. Efforts in creating artificial mimics of antibodies have been a significant focus in chemical biology and bioorganic studies. Among various molecules of interest, two types of compounds have particularly secured spotlights: DNA aptamers and supramolecular peptides. Both of them have demonstrated specific recognition toward target molecules in a plethora of studies. Although fairly well established, these two versatile platforms continue to be employed for new challenges. In the past few years, many novel aptamer- and peptide-based analytical methods have been developed to achieve cancer detection and tumor analysis.

The concept of using aptamers as antibody mimics has been around for quite a while. Complete assemblies such as an enzyme-linked aptamer assay have also been demonstrated in many studies. These sandwich-type assays use one or several aptamers to capture and analyze proteins, similar to enzyme-linked immunosorbent assays (ELISA). Recent development mainly concerns improving detection limit as well as establishing new targets. Kim et al. developed such a DNA aptamer-based assay for detecting lipocalin-2 (LCN2) in hepatocellular carcinoma. This sandwich assay platform contains three major steps: the functionalization of surface with capture aptamer, hybridization/recognization, and signal read out by HRP-labeled reporter aptamer. The device can sensitively detect LCN2 over the range of 2.5–500 ng/mL with a detection limit of 0.6 ng/mL. The system shows a broader detection range and comparable sensitivity when compared with the performance of the commercially available antibody-based ELISA kits (Lee et al., 2015).

Peptides that can selectively bind to proteins can also be used for bioanalysis. Heath et al. developed a general epitope-targeting strategy to screen peptide-based ligands (Das et al., 2015). Through this strategy, they discovered a selective peptide ligand targeting the E17K point mutation in the pleckstrin homology domain of the Akt1 oncoprotein. The ligand exhibits 10 times higher selectivity for the oncoprotein compared to the wild type. The authors demonstrated that the Cy5-labeled ligand is preferentially retained in cells with E17K mutation of Akt1 rather than in the wild-type cells (Deyle et al., 2015).

Inorganic supramolecular sensors have also recently been employed in direct supramolecular recognition-based analytical methods. Notably, Zaworotko et al. constructed two lanthanide zeolite-like metal-organic frameworks (Ln-ZMOFs) by self-assembly of a 4-connected lanthanide molecular building block and a bipyridine–dicarboxylate ligand. Adjusting the ratio of Tb^{3+} and Eu^{3+} yielded three mixed-crystal isostructural MZMOFs, all of which exhibit the characteristic emission bands of both Tb^{3+} and Eu^{3+} ions. With the addition of lysophosphatidic acid (LPA, a biomarker of ovarian cancer), the emission of Tb^{3+} ion decreases whereas the emission of Eu^{3+} increases. The intensity ratio of Eu^{3+} and Tb^{3+} correlated well with the concentration of LPA. Thereinto, one of them, $Eu_{0.6059}Tb_{0.3941}$–ZMOF, shows the highest selectivity toward LPA. Quantitative measurement shows that LPA can be sensitively detected in the physiological concentration range of 1.4–43.3 μM (Zhang, Shi, Cheng, & Zaworotko, 2015).

6. CONCLUDING REMARKS AND OUTLOOK

The field of supramolecular analytical chemistry has always been interested in cancer research, evidenced by its continuous supply of novel detection schemes. There has not been a sign of slowing down in this opportunity-filled land: novel analytical techniques with fascinating supramolecular chemistry are published on a daily basis. Despite the prosperity in academic studies, most of the methods are far from eliciting clinical influences. Comparing to the clinical progress made in the supramolecular drug delivery field, the contribution from supramolecular analytical chemistry seems rather deficient. Although it is exciting to see that more and more studies have started to involve patient samples, we do believe that even more efforts should be devoted to implementing these technologies in a clinical oncology setting. Perhaps in the near future, supramolecular analytical techniques may become as indispensable as nature's antibodies.

REFERENCES

Alhadrami, H. A., Chinnappan, R., Eissa, S., Rahamn, A. A., & Zourob, M. (2017). High affinity truncated DNA aptamers for the development of fluorescence based progesterone biosensors. *Analytical Biochemistry*, *525*, 78–84.

Anslyn, E. V. (2007). Supramolecular analytical chemistry. *The Journal of Organic Chemistry*, *72*(3), 687–699.

Cheng, C., McGonigal, P. R., Schneebeli, S. T., Li, H., Vermeulen, N. A., Ke, C., et al. (2015). An artificial molecular pump. *Nature Nanotechnology*, *10*(6), 547–553.

Chinen, A. B., Guan, C. M., Ferrer, J. R., Barnaby, S. N., Merkel, T. J., & Mirkin, C. A. (2015). Nanoparticle probes for the detection of cancer biomarkers, cells, and tissues by fluorescence. *Chemical Reviews*, *115*(19), 10530–10574.

Cragg, P. J. (2010). *Supramolecular chemistry: From biological inspiration to biomedical applications*. Springer Science & Business Media.

Das, P., & Krull, U. J. (2017). Detection of a cancer biomarker protein on modified cellulose paper by fluorescence using aptamer-linked quantum dots. *Analyst*, *142*(17), 3132–3135.
Das, S., Nag, A., Liang, J., Bunck, D. N., Umeda, A., Farrow, B., et al. (2015). A general synthetic approach for designing epitope targeted macrocyclic peptide ligands. *Angewandte Chemie International Edition*, *54*(45), 13219–13224.
Davis, M. E. (2009). The first targeted delivery of siRNA in humans via a self-assembling, cyclodextrin polymer-based nanoparticle: From concept to clinic. *Molecular Pharmaceutics*, *6*(3), 659–668.
Davis, M. E., Chen, Z., & Shin, D. M. (2008). Nanoparticle therapeutics: An emerging treatment modality for cancer. *Nature Reviews Drug Discovery*, *7*, 771–782.
Deyle, K. M., Farrow, B., Qiao Hee, Y., Work, J., Wong, M., Lai, B., et al. (2015). A protein-targeting strategy used to develop a selective inhibitor of the E17K point mutation in the pH domain of Akt1. *Nature Chemistry*, *7*, 455–462.
Dong, L., Miao, Q., Hai, Z., Yuan, Y., & Liang, G. (2015). Enzymatic hydrogelation-induced fluorescence turn-off for sensing alkaline phosphatase in vitro and in living cells. *Analytical Chemistry*, *87*(13), 6475–6478.
Esteller, M., & Pandolfi, P. P. (2017). The epitranscriptome of noncoding RNAs in cancer. *Cancer Discovery*, *7*(4), 359–368.
Golubkov, V. S., Prigozhina, N. L., Zhang, Y., Stoletov, K., Lewis, J. D., Schwartz, P. E., et al. (2014). Protein–tyrosine pseudokinase 7 (PTK7) directs cancer cell motility and metastasis. *Journal of Biological Chemistry*, *289*(35), 24238–24249.
Han, A., Wang, H., Kwok, R. T. K., Ji, S., Li, J., Kong, D., et al. (2016). Peptide-induced AIEgen self-assembly: A new strategy to realize highly sensitive fluorescent light-up probes. *Analytical Chemistry*, *88*(7), 3872–3878.
He, X.-P., Hu, X.-L., Jin, H.-Y., Gan, J., Zhu, H., Li, J., et al. (2015). Quick serological detection of a cancer biomarker with an agglutinated supramolecular glycoprobe. *Analytical Chemistry*, *87*(17), 9078–9083.
Hou, T.-C., Wu, Y.-Y., Chiang, P.-Y., & Tan, K.-T. (2015). Near-infrared fluorescence activation probes based on disassembly-induced emission cyanine dye. *Chemical Science*, *6*(8), 4643–4649.
Hu, Y., Zhang, L., Zhang, Y., Wang, B., Wang, Y., Fan, Q., et al. (2015). Plasmonic nanobiosensor based on hairpin DNA for detection of trace oligonucleotides biomarker in cancers. *ACS Applied Materials & Interfaces*, *7*(4), 2459–2466.
Kataoka, K., Harada, A., & Nagasaki, Y. (2012). Block copolymer micelles for drug delivery: Design, characterization and biological significance. *Advanced Drug Delivery Reviews*, *64*, 37–48.
Kumar, M., Brocorens, P., Tonnelé, C., Beljonne, D., Surin, M., & George, S. J. (2014). A dynamic supramolecular polymer with stimuli-responsive handedness for in situ probing of enzymatic ATP hydrolysis. *Nature Communications*, *5*, 5793.
Kwok, R. T. K., Leung, C. W. T., Lam, J. W. Y., & Tang, B. Z. (2015). Biosensing by luminogens with aggregation-induced emission characteristics. *Chemical Society Reviews*, *44*(13), 4228–4238.
Le, N. D. B., Yesilbag Tonga, G., Mout, R., Kim, S.-T., Wille, M. E., Rana, S., et al. (2017). Cancer cell discrimination using host–guest "doubled" arrays. *Journal of the American Chemical Society*, *139*(23), 8008–8012.
Lee, K.-A., Ahn, J.-Y., Lee, S.-H., Singh Sekhon, S., Kim, D.-G., Min, J., et al. (2015). Aptamer-based sandwich assay and its clinical outlooks for detecting lipocalin-2 in hepatocellular carcinoma (HCC). *Scientific Reports*, *5*(1).
Li, Y., Sun, L., & Zhao, Q. (2017). Competitive fluorescence anisotropy/polarization assay for ATP using aptamer as affinity ligand and dye-labeled ATP as fluorescence tracer. *Talanta*, *174*, 7–13.

Li, S., Xu, L., Ma, W., Wu, X., Sun, M., Kuang, H., et al. (2016). Dual-mode ultrasensitive quantification of microRNA in living cells by chiroplasmonic nanopyramids self-assembled from gold and upconversion nanoparticles. *Journal of the American Chemical Society, 138*(1), 306–312.

Lin, Y., Chapman, R., & Stevens, M. M. (2014). Label-free multimodal protease detection based on protein/perylene dye coassembly and enzyme-triggered disassembly. *Analytical Chemistry, 86*(13), 6410–6417.

Lin, S., Gao, W., Tian, Z., Yang, C., Lu, L., Mergny, J.-L., et al. (2015). Luminescence switch-on detection of protein tyrosine kinase-7 using a G-quadruplex-selective probe. *Chemical Science, 6*(7), 4284–4290.

Liu, Y., Niu, L.-Y., Chen, Y.-Z., & Yang, Q.-Z. (2018). A self-assembled fluorescent nanoprobe for detection of gsh and dual-channel imaging. *Journal of Photochemistry and Photobiology A: Chemistry, 355*, 311–317.

Ma, X., & Zhao, Y. (2015). Biomedical applications of supramolecular systems based on host–guest interactions. *Chemical Reviews, 115*(15), 7794–7839.

Mei, J., Leung, N. L. C., Kwok, R. T. K., Lam, J. W. Y., & Tang, B. Z. (2015). Aggregation-induced emission: Together we shine, united we soar! *Chemical Reviews, 115*(21), 11718–11940.

Panda, D., Debnath, M., Mandal, S., Bessi, I., Schwalbe, H., & Dash, J. (2015). A nucleus-imaging probe that selectively stabilizes a minor conformation of c-MYC G-quadruplex and down-regulates c-MYC transcription in human cancer cells. *Scientific Reports, 5*(1).

Peer, D., Karp, J. M., Hong, S., Farokhzad, O. C., Margalit, R., & Langer, R. (2007). Nanocarriers as an emerging platform for cancer therapy. *Nature Nanotechnology, 2*, 751–760.

Pittolo, S., Gómez-Santacana, X., Eckelt, K., Rovira, X., Dalton, J., Goudet, C., et al. (2014). An allosteric modulator to control endogenous g protein-coupled receptors with light. *Nature Chemical Biology, 10*, 813–815.

Ren, C., Zhang, J., Chen, M., & Yang, Z. (2014). Self-assembling small molecules for the detection of important analytes. *Chemical Society Reviews, 43*(21), 7257–7266.

Ribas, A., Kalinoski, L., Heidel, J. D., Peterkin, J., Seligson, D. B., Zuckerman, J. E., et al. (2010). Systemic delivery of siRNA via targeted nanoparticles in patients with cancer: Results from a first-in-class phase I clinical trial. Journal of Clinical Oncology, 28(15 suppl), 3022.

Sun, Q., Fang, S., Fang, Y., Qian, Z., & Feng, H. (2017). Fluorometric detection of cholesterol based on β-cyclodextrin functionalized carbon quantum dots via competitive host–guest recognition. *Talanta, 167*, 513–519.

van Dongen, S. F. M., Cantekin, S., Elemans, J. A. A. W., Rowan, A. E., & Nolte, R. J. M. (2014). Functional interlocked systems. *Chemical Society Reviews, 43*(1), 99–122.

Wang, K., Cui, J.-H., Xing, S.-Y., & Dou, H.-X. (2016). A calixpyridinium-based supramolecular tandem assay for alkaline phosphatase and its application to ATP hydrolysis reaction. *Organic & Biomolecular Chemistry, 14*(9), 2684–2690.

Wang, M., Mao, Z., Kang, T.-S., Wong, C.-Y., Mergny, J.-L., Leung, C.-H., et al. (2016). Conjugating a groove-binding motif to an Ir(III) complex for the enhancement of G-quadruplex probe behavior. *Chemical Science*, 7(4), 2516–2523.

Wei, P., Yan, X., & Huang, F. (2015). Supramolecular polymers constructed by orthogonal self-assembly based on host–guest and metal–ligand interactions. *Chemical Society Reviews, 44*(3), 815–832.

Xue, M., Wei, W., Su, Y., Johnson, D., & Heath, J. R. (2016). Supramolecular probes for assessing glutamine uptake enable semi-quantitative metabolic models in single cells. *Journal of the American Chemical Society, 138*(9), 3085–3093.

Xue, M., Wei, W., Su, Y., Kim, J., Shin, Y. S., Mai, W. X., et al. (2015). Chemical methods for the simultaneous quantitation of metabolites and proteins from single cells. *Journal of the American Chemical Society, 137*(12), 4066–4069.

Xue, M., & Zink, J. I. (2013). An enzymatic chemical amplifier based on mechanized nanoparticles. *Journal of the American Chemical Society*, *135*(47), 17659–17662.
Yang, T., Cheong, A., Mai, X., Zou, S., & Woon, E. C. Y. (2016). A methylation-switchable conformational probe for the sensitive and selective detection of RNA demethylase activity. *Chemical Communications*, *52*(36), 6181–6184.
Yoshii, T., Mizusawa, K., Takaoka, Y., & Hamachi, I. (2014). Intracellular protein-responsive supramolecules: Protein sensing and in-cell construction of inhibitor assay system. *Journal of the American Chemical Society*, *136*(47), 16635–16642.
You, L., Zha, D., & Anslyn, E. V. (2015). Recent advances in supramolecular analytical chemistry using optical sensing. *Chemical Reviews*, *115*(15), 7840–7892.
Yuan, Y., Zhang, J., Cao, Q., An, L., & Liang, G. (2015). Intracellular disassembly of self-quenched nanoparticles turns NIR fluorescence on for sensing furin activity in cells and in tumors. *Analytical Chemistry*, *87*(12), 6180–6185.
Zhang, J., & Ma, P. X. (2013). Cyclodextrin-based supramolecular systems for drug delivery: Recent progress and future perspective. *Advanced Drug Delivery Reviews*, *65*(9), 1215–1233.
Zhang, S.-Y., Shi, W., Cheng, P., & Zaworotko, M. J. (2015). A mixed-crystal lanthanide zeolite-like metal-organic framework as a fluorescent indicator for lysophosphatidic acid, a cancer biomarker. *Journal of the American Chemical Society*, *137*(38), 12203–12206.
Zhang, S., Zhao, B. S., Zhou, A., Lin, K., Zheng, S., Lu, Z., et al. (2017). m6A demethylase ALKBH5 maintains tumorigenicity of glioblastoma stem-like cells by sustaining FOXM1 expression and cell proliferation program. *Cancer Cell*, *31*(4), 591–606, e596.
Zhao, B., Wu, P., Zhang, H., & Cai, C. (2015). Designing activatable aptamer probes for simultaneous detection of multiple tumor-related proteins in living cancer cells. *Biosensors and Bioelectronics*, *68*, 763–770.
Zhu, Y., Yang, B., Chen, S., & Du, J. (2017). Polymer vesicles: Mechanism, preparation, application, and responsive behavior. *Progress in Polymer Science*, *64*, 1–22.

CHAPTER SEVEN

Gold Nanoparticles for the Delivery of Cancer Therapeutics

Dean M. Connor[*,†], **Ann-Marie Broome**[*,†,‡,1]

[*]Department of Cell and Molecular Pharmacology & Experimental Therapeutics, Medical University of South Carolina, Charleston, SC, United States

[†]Hollings Cancer Center, Medical University of South Carolina, Charleston, SC, United States

[‡]Department of Neurosciences, Medical University of South Carolina, Charleston, SC, United States

[1]Corresponding author: e-mail address: broomea@musc.edu

Contents

Abstract

Gold nanoparticles (Au NPs) are very attractive and versatile nanoparticles since they have a remarkable capacity to absorb and scatter light, convert optical energy into heat via nonradiative electron relaxation dynamics, and surface chemistries that can be capitalized upon so that the nanoparticles act as drug carriers. Au NPs have excellent stability and biocompatibility, tailorable shapes and sizes, an easily functionalized surface, high drug-loading capacity, and low toxicity. The properties of Au NPs can be leveraged to develop more precisely targeted and effective cancer therapeutics. Au NPs have been used to target delivery of chemotherapeutic agents, complement radiation and thermal therapy, and enhance contrast for in vivo imaging of the tumor in a variety of cancer types and diseased organs.

Advances in Cancer Research, Volume 139
ISSN 0065-230X
https://doi.org/10.1016/bs.acr.2018.05.001

1. INTRODUCTION

Nanotechnology takes a multidisciplinary approach to therapy with broad potential applications in cancer treatment. Continued development of cancer nanotechnologies holds promise for oncologic therapies in which nanoparticles (NPs) can be used to diagnose and treat cancer in a patient. Targeting ligands, imaging labels, therapeutic drugs, and other functionalities can be integrated on or within NPs to allow for optimal functionality.

NPs administered either systemically or locally can transport therapeutic drugs to a target organ or area. NPs coated with biocompatible polymers utilize immune system-evading behavior which can postpone or prevent the clearance by the reticuloendothelial system (RES) and prolong circulation time in the blood (Gref et al., 2000). Due to tumor vessel permeability and the poor lymphatic recovery system in the tumor, nanosized particles administered parenterally can extravasate through the endothelium and accumulate in solid tumors by the enhanced permeability and retention (EPR) effect (Maeda, 2001; Torchilin, 2006). Although NPs can passively accumulate in solid tumors, the release and delivery of drugs deep into the bulk of the tumors remain challenging. Important goals of NP-based drug delivery systems are to: transport therapeutic drugs with poor water solubility, provide a reasonably long circulation time, improve drug biodistribution, enable efficient release, minimize side effects, and allow for degradation or excretion.

In particular, gold NPs (Au NPs) are very attractive and versatile NPs since they have a remarkable capacity to absorb and scatter light, convert optical energy into heat via nonradiative electron relaxation dynamics, and surface chemistries that can be capitalized upon so that the NPs act as drug carriers. Au NPs have excellent stability and biocompatibility, tailorable shapes and sizes, an easily functionalized surface, high drug-loading capacity, and low toxicity (Boisselier & Astruc, 2009). The NP acts as a water-soluble and biocompatible platform that allows the delivery of drugs to the site of therapy. Due to favorable surface area characteristics, hundreds of molecules can be attached to a single Au NP. Au NPs can also be labeled to target specific cancer cell ligands and therefore can be used to target only malignant cells. Therefore, Au NPs can be both theranostic and curative in nature. In this chapter, we will establish the clinical need for Au NPs in cancer therapeutics. We will then review the properties and traits of Au NPs that are relevant to addressing the clinical need.

2. FUNDAMENTAL PROPERTIES OF GOLD NANOPARTICLES

In order to understand how Au NPs can effectively be added into a basic science or clinical research protocol, we will begin by reviewing the properties of Au NPs and how those properties can help us meet the clinical need for more targeted delivery of cancer therapeutics. Size, shape, charge, toxicity, and biocompatibility of Au NPs will be examined.

2.1 Size of Gold Nanoparticles

In general, NPs are man-made particles that range in size from a few nanometers to about a few hundred nanometers along the narrowest two dimensions. The size of Au NPs can have several, possibly competing, effects on their use in cancer therapeutics. The size of the NP determines its physical characteristics and its surface area, such as the color of the NP, and it even effects how the body interacts with the NP.

One of the primary reasons to consider using NPs in cancer therapeutics is the ability to functionalize the Au NP and selectively deliver a drug only to the site of the tumor. Both functionalizing the Au NP and delivering a drug require that the drug and functionalizing agent attach to the surface of, or surface layer over, the Au NP. Therefore, to the extent that it is feasible, maximizing the available surface area is advantageous. For spherical NPs, the ratio of surface area to volume increases as the radius decreases, so to maximize total available surface area the preference is for more, smaller nanoparticles over fewer, larger NPs. There are other considerations such as toxicity and biocompatibility that will need to be weighed against the advantages of increased surface area with decreased NP size.

When NPs begin approaching the hundreds-of-nanometers scale, their physical properties begin to transform and seemingly evade the laws of physics. For Au NPs, the surface electrons resonate (slosh back and forth) at a frequency determined by the size of the NP—this phenomenon, called surface plasmon resonance, is characteristic of the size of the Au NP and has both a characteristic reflection wavelength and transmission wavelength. A classic example of this is the Lycurgus Cup which was made in the 4th century. When lit from the front (so the viewer sees the reflected light) the cup appears to be green; when lit from behind (so the viewer sees the transmitted light), the cup appears red (Merali, 2013). In addition, while the color of bulk gold is well known, suspended solutions of spherical Au NPs can vary in color from clear (~2 nm) to red (~30 nm) to deep red to purple/blue (~100 nm).

For uncoated Au NPs (colloidal gold), the size of the NP will directly affect how the body processes or treats the nanoparticles. For instance, NPs smaller than 100 nm can easily enter most cells. Those NPs smaller than 20 nm can indiscriminately transit out of blood vessels. Au NPs in the range of 20–100 nm are generally considered to be the most safe, but the combination of NP size, shape, and surface coating must be considered as well (Goodman, McCusker, Yilmaz, & Rotello, 2004; Lewinski, Colvin, & Drezek, 2008; Tkachenko et al., 2004), and toxicity/biocompatibility studies need to be performed on any new size/shape/surface-coating combinations.

2.2 Shape of Gold Nanoparticles

Au NPs can be fabricated in shapes from solid spheres, shells, cages, rods, chains (Karathanasis & Ghaghada, 2016), and stars. As with the size of NPs, the shape also affects the surface area, i.e., how the body interacts with the NP, and the color of the NP. The shapes that are currently of most interest are solid spheres, nanorods (Au NPs that are narrow in two dimensions and longer, ~100 nm, along a third dimension), and nanostars (highly symmetric (Niu, Chua, Zhang, Huang, & Lu, 2015), star-shaped Au NPs).

For spherical Au NPs, since the NPs are symmetrical in three dimensions, the surface plasmon resonance is single valued and the light absorption, reflection, and transmission behavior of the Au NPs are single valued as well. Since gold nanorods are symmetrical about two axes, but longer along the third axis, gold nanorods have bimodal (transverse and longitudinal) surface plasmon resonance behaviors—the longitudinal mode of the gold nanorod corresponds to a longer wavelength for absorption of light by the NP. Generally, Au NPs are designed to have an absorption peak (from the longitudinal mode) in the near-infrared (Zhang et al., 2018). Near-infrared wavelengths penetrate much farther into tissue than visible light and are used for photothermal therapy (Huang et al., 2011; Murphy et al., 2008; Zhang et al., 2018).

2.3 Charge of Gold Nanoparticles

To keep Au NPs from aggregating in solution, they need to carry a net positive or negative electrical charge, thus repelling each other but yet remaining soluble. The surface charge on the NP is described by its zeta potential (a measure of the electrical potential at the surface of the NPs relative to a distant, charge-free region). Uncoated Au NPs need to have a zeta potential of at least ±20 mV to stay stable in the solution (Conde et al., 2012). The zeta potential of the Au NPs needs to be carefully selected on a case-by-case basis, based on surface-coating parameters. If one wants a positively charged molecule (or the positive end of a dipole molecule) to bond with the Au NP, then Au NPs with a negative potential must be used to attract the positive molecules, or positive ends of a dipolar molecule.

2.4 Gold Nanoparticles as Drug Carriers

Chemotherapeutics function by causing cell death or slowing the proliferation and spread of cancer cells. In general, chemotherapeutics directly damage the genetic code and cellular machinery within the nucleus or damage chemical processes within mitosis and cell cycling, thus preventing the cell from dividing uncontrollably. This can be accomplished by inhibiting microtubule function, protein function, or DNA synthesis. Several categories of chemotherapeutics exist based on the goals of the therapy, including curative chemotherapy (eliminate the cancer cells from the body), adjuvant chemotherapy (elimination of residual cancer cells after surgery or to prevent recurrence), neoadjuvant chemotherapy (shrinkage of tumors prior to surgery), and palliative chemotherapy (relieve certain symptoms or temporarily slow cancer progress). Linking chemotherapeutics to an Au NP can enhance the functionality of the drug by delivering the complex undeterred to a designated site of action.

As a drug carrier, Au NPs can transport therapeutic agents through either covalent bonding or noncovalent attachment (Cheng et al., 2008; Friedman, Kerby, & Calvert, 2000; Ho et al., 2014; Marko et al., 2014; Nava et al., 2014; Noone et al., 2018; Stupp et al., 2005). While covalent bonding requires chemical modification of the drug and external triggers for drug release, noncovalent adsorption of active drug on Au NPs provides an alternative way to transport and efficiently release the drug (Cheng et al., 2008; Marko et al., 2014; Stupp et al., 2005). It has been demonstrated that the polyethylene glycol (PEG)-coated Au NP with an NP core size ≤5 nm can efficiently transport active hydrophobic drugs into cancer cells mainly by membrane diffusion (Friedman et al., 2000; Marko et al., 2014; Stupp et al., 2005). The fate of drug molecules delivered by Au NPs in cells is in some cases well studied, but there is insufficient information about the fate of Au NP entrapped drug obtained in animal models.

2.5 Toxicity and Biocompatibility of Gold Nanoparticles

For Au NPs to be safe and efficient drug delivery vehicles, they should not be cytotoxic and should not affect cellular physiology as this could impede normal cell functionality or lead to undesired side effects. Assessing the biocompatibility of the nanoparticles has proven to be quite a difficult task. Problems inherent with Au NP delivery include pH changes, circulating enzymes, rapid transit away from absorptive sites, and poor absorption. Potential toxicity stems from a balance of surface chemistry, coating materials, size, shape, and target. Research suggests that naked Au NPs (i.e., as synthesized) are extremely toxic, but with surface modification (discussed later) the toxicity can be ameliorated. For example, positive Au NPs are more likely to be toxic than negatively charged or neutral NPs based on the ability of the NPs to interact with and disrupt cellular membranes (Harper et al., 2011; Schaeublin et al., 2011). In addition, Au NPs can have a direct impact on cellular morphology, mitochondrial function, mitochondrial membrane potential (MMP), intracellular calcium levels, DNA damage-related gene expression, and p53 and caspase-3 expression. To reduce the apparent toxicity of the Au NPs, their surfaces must be modified.

2.6 Surface Modification of Gold Nanoparticles

Targeted drug delivery can be achieved by modification of the Au NP surface. A significant number of available, effective drugs have been modified to covalently bind to the Au NPs (Gibson, Khanal, & Zubarev, 2007; Hone et al., 2002). Even though the modified prodrugs can be stabilized by a chemical bond on the NP surface, the efficient release and efficacy of the drugs are reduced. Only a few examples that employed noncovalent adsorption of drugs to Au NPs have demonstrated rapid release of the bound drug without use of external stimuli (Cheng et al., 2008, 2010; Ito et al., 2009; Oo, Yang, Du, & Wang, 2008). Au NPs coated with PEG are used as noncovalent drug delivery systems with unique advantages for systemic administration (Cheng et al., 2008). The Au NP surface provides an amphiphilic environment for lipophilic drugs. In addition, the PEG on the Au surface renders the nanoparticle perfectly soluble in water and minimizes any protein adsorption (Cheng et al., 2011). In addition, it has been widely reported that PEGylated NPs have "stealth" character, which can postpone or prevent the rapid clearance by the RES (Cai et al., 2018). Therefore, drugs on these Au NP carriers can have a programmed circulation time in the blood.

Targeted strategies have several advantages over systemic strategies: lower doses required, greater control over toxicity and bioavailability of dose, extended duration of release, possibilities to combine local and systemic drugs with different kinetics, controlled release directly to site, and avoidance of systemic drug exposure. Several different targeting approaches have been implemented. An example of such an approach is to conjugate the Au NP with HER1 or HER2 monoclonal antibodies that bind to the epidermal growth factor receptor (EGFR), which is significantly upregulated in many cancer cells (Johnston et al., 2006; Xiong et al., 2004). Another example of such a targeting approach is to utilize peptides or full-length ligands. A classic example involves RGD-containing peptides which target $\alpha_V\beta3$ or $\alpha_V\beta5$ integrin. This is particularly beneficial in targeting tumor angiogenesis (Haubner et al., 2001).

3. ROLE OF GOLD NANOPARTICLES IN TREATING SPECIFIC CANCERS

In this section, we will break down by organ system recent advances in the use of Au NPs in treating different cancer types. Although this is not a comprehensive treatment of all organ systems or cancer types, it will give a taste for the ways in which the properties of Au NPs can be leveraged to develop more precisely targeted and more effective cancer therapeuics. The approaches of using Au NPs tend to fall into one of three, often complementary, categories: targeted delivery of a chemotherapeutic agent, enhancement of radiation/thermal therapy, and enhancement of contrast for in vivo imaging of the tumor.

3.1 Reproductive Organ Cancers

Currently, Au NPs have been studied extensively in treating prostate and breast cancer, two of the most common types of cancer in men and women, respectively. The earliest approaches to using Au NPs for improving cancer treatment efficacy leveraged the EPR effect to preferentially deliver colloidal Au NPs to vicinity of the tumor, thereby increasing the local attenuation of gamma ray treatment beam of radiation therapy and ultimately delivering an enhanced radiation dose to the tumor site. While this approach has proven to be effective enhancing radiation dose in specific instances, the consensus of the medical physics community is that the wide variety of size, shape, charge, and coating make it challenging to make general statements of the efficacy of Au NPs for enhancing radiation therapy dose (Schuemann et al., 2016). In addition, a more targeted approach that maximizes the uptake of Au NPs into the tumor cell nuclei is needed to maximize the radiation dose enhancement effect of Au NPs. In one promising recent study, Wang et al. demonstrated that glucose-capped Au NPs led to a significant enhancement of radiation-dose-based cytotoxicity (radiosensitivity) to MDA-MB-231 breast cancer by Au NPs (Wang, Jiang, Li, & Hu, 2015). They showed that the glucose-capped Au NPs would uptake into the cytoplasm of the tumor cells, but stopped short of entering the cell nucleus. Even with the subideal uptake of Au NPs, they were able to demonstrate an 86% improvement in radiosensitivity over controls (Wang et al., 2015). Similarly, Wu et al. have shown that RGD-conjugated Au NPs yield enhanced radiosensitivity for breast cancer cells (Wu et al., 2017). Wolfe et al. showed that goserelin-conjugated gold nanorods could be used to enhance the radiosensitivity of PC3 prostate cancer cells in mice (Wolfe et al., 2015). As with the previous example, the goserelin-conjugated gold nanorods were localized into the cytoplasm of the PC3 cells, but still led to a 36% enhancement in the radiosensitivity of the PC3 cells.

The majority of the studies on the efficacy of Au NP-based approaches to cancer therapeutics in reproductive cancers are focused on the targeted delivery of chemotherapeutic drugs. In these studies, Au NPs are typically conjugated with a targeting agent and loaded with at least one chemotherapeutic drug. For example, Au NPs were conjugated with a prostate-specific membrane antigen (PSMA) RNA antigen to bind to PSMA in prostate cancer cells and loaded with doxorubicin (Kim, Jeong, & Jon, 2010). This combination of Au NP, targeting antigen, and chemotherapeutic was shown to enhance both the contrast of the tumor in CT imaging and the efficacy of the chemotherapeutic agent for LNCaP and PC3 prostate cancer cell lines (Gautier, Allard-Vannier, Munnier, Soucé, & Chourpa, 2013; Kim, Jeong, & Jon, 2010). Murawala et al. demonstrated that methotrexate (MTX)-loaded, bovine serum albumin-capped Au NPs led to drastic enhancement of tumoricidal effect over the MTX or Au NPs alone (Murawala, Tirmale, Shiras, & Prasad, 2014).

Several recent studies have aimed to maximize the effectiveness of Au NP-based chemotherapeutics through simultaneously targeting specific tumor types, loading the nanoparticle with a chemotherapeutic, and using optimized gold nanorods for photothermal therapy (Banu et al., 2015; Ju et al., 2017; Mendes, Pedrosa, Lima, Fernandes, & Baptista, 2017; Peralta, Heidari, Dash, & Tarr, 2015; Wang et al., 2014).

3.2 Central Nervous System Cancers

Brain tumors (gliomas) are among the most deadly tumor subtypes because tumor location, grade, histopathology, and recurrence play critical roles in the patient's survival rate. Poor accessibility of therapeutic drugs into the central nervous system (CNS) is primarily due to the blood–brain barrier (BBB), which limits the treatment options for neurological disorders, such as brain tumors. The BBB is a specialized semipermeable membrane barrier composed of capillary endothelial cells that line the cerebral microvessels and surrounding perivascular elements. The unique tight junctions between the endothelial cells are very efficient at restricting access to the CNS. The BBB structure only allows for passive diffusion of gases, water, and hydrophobic molecules with a molecular weight less than 400–500 Da. Other nutrients and endogenous molecules, such as amino acids, glucose, vitamins, or essential fatty acids, must be actively transported across the membrane by transporters found at the BBB. In addition, the BBB is a metabolic barrier as it contains enzymes that metabolize and inactivate toxins, neuroactive compounds, and chemotherapeutics.

The majority of NP vehicles use either passive absorption or receptor-mediated transcytosis to cross the BBB. Incorporating a targeting moiety, i.e., peptide or peptidomimetic monoclonal antibody, that is, recognized by receptors present at the BBB endothelium is the most common strategy to evade the BBB. As a result, Au NPs have emerged as an alternative delivery vehicle to the CNS due to their small size and surface properties. Several endogenous BBB receptors have been utilized as targets with modified ligands as the targeting moieties, such as transferrin (Tf) (Cai et al., 2018; Dixit et al., 2015), insulin, glutathione (Yu et al., 2016), and low-density lipoprotein (LDL) (Demeule et al., 2008).

A major clinical opportunity for Au NPs is to deliver drugs to glioblastoma multiform (GBM), the single deadliest primary brain tumor. GBMs are also the most common primary brain tumor in adults with approximately 12,000 new diagnoses a year in the United States (Ho et al., 2014; Nava et al., 2014). GBMs are difficult tumors to treat with a 5-year survival rate of <5% (Noone et al., 2018), even with aggressive therapy including surgery, chemotherapy, and radiation (Marko et al., 2014). Surgical resection to remove the tumor is the mainstay of treatment. In most cases, curative resection is not possible due to infiltrating growth of the tumor into normal brain parenchyma. Standard of care (SOC) systemic therapy prolongs survival for mere months compared with no treatment (Stupp et al., 2005).

Various nanoparticles have been used in the past to deliver drugs and image brain gliomas across the BBB. A few examples include super paramagnetic iron oxide particles to target and image gliomas in a rat model (Jiang et al., 2012) and PLGA nanoparticles (Cui, Xu, Chow, Wang, & Wang, 2013) coated with Tf and loaded with doxorubicin and paclitaxel were studied for tumor inhibition. A major concern with approaches using these nanocarriers is the significant risk of immunological responses associated with long-term treatment. In addition, the fate of these agents and their degradation by-products can have a long-term effect on healthy tissues.

In contrast, Au NPs less than 8 nm in diameter have been successfully targeted to GBM using receptor ligands. For instance, Tf-coated Au NPs have been shown to specifically target tumor cells (Dixit et al., 2015). A multifunctional Tf-coated Au NP that encapsulates hydrophobic drugs via a noncovalent linkage is preferable. Noncovalent linkage does not require modifications to FDA-approved drug compositions and the drug is maintained in its active formulation. Noncovalent attachment of the drug also facilitates ease of loading and offloading of the drug as compared to covalent linkages or modifications of the drug used in the past. Cheng et al. (2008) previously demonstrated noncovalent drug loading on Au NPs for highly efficient drug delivery in cancer cells. The drug delivery time was significantly reduced to 2 h instead of days for the free drug due to this noncovalent attachment. Targeting of Au NPs with EGF peptide also reduced systemic cytotoxicity and increased drug delivery efficiency as compared to untargeted Au NPs (Cheng et al., 2011).

3.3 Digestive System Cancers

Since one of the liver's primary functions is to filter chemicals from the blood stream, at first glance it would seem that liver cancer would be an ideal candidate for targeted Au NPs, but using Au NPs to treat liver cancer, and, similarly, spleen cancer, presents a particular challenge (Zhang, Poon, Tavares, McGilvray, & Chan, 2016). The challenge is to design targeted Au NPs that will not get caught by one of the liver's many mechanisms for filtering the blood. Targeted Au NPs that are filtered out by the liver will not localize to the tumor site and will negate the advantage of using targeted nanoparticles for cancer therapeutics. In Yan's review of approaches to active targeting nanoparticles to hepatocellular carcinoma (HCC), they narrowed the best candidates to transferrin receptor, EGFR, CD44, and integrin for active and specific targeting to HCC (Yan, Liao, Lin, & He, 2015).

While these candidates are a good starting point, thus far, few studies have simultaneously demonstrated the ability of Au NPs to bypass hepatocytic uptake and to specifically target liver tumor cells.

4. ASSESSMENT OF CANCER THERAPEUTICS DELIVERY AND EFFICACY

The deepest clinical need in—and the holy grail of—cancer therapeutics is a drug that will travel the human body, find all the cancer cells of the primary and metastatic tumors, and selectively kill the tumor cells while sparing all healthy tissue. We have established that, in some cases, Au NP-based cancer therapeutics can meet this clinical need, but how do we know? How do we know how much of a drug is delivered to the tumor site (vs how much is delivered to healthy tissue)? How do we know if the drug is effectively killing the tumor?

In this section, we will establish the use of in vivo preclinical and clinical imaging for quantifying the delivery of Au NP-based cancer therapeutics to the tumor site and show how these same imaging approaches can be used to assess the efficacy of the therapeutics. We will break down our treatment into two broad (sometimes overlapping) categories: preclinical imaging (typically imaging of small animals) and clinical imaging (typically screening or diagnostic imaging of human cancer patients). While there are broader considerations for each of these modalities, we will specifically focus on each imaging system's ability to quantify the uptake of Au NPs tagged with chemotherapeutics and ability to assess the efficacy of the chemotherapeutic.

4.1 Preclinical Imaging

While the field of preclinical imaging is broad, we will focus specifically on in vivo preclinical imaging in this chapter. Specifically, we will look at four of the most commonly used preclinical imaging modalities (bioluminescence imaging, fluorescence imaging, microcomputed tomography (micro-CT), and micro-positron-emission tomography (micro-PET)).

4.1.1 Bioluminescence

Bioluminescence is the workhorse imaging modality of most preclinical imaging regimens. Bioluminescence is the emission of energy from a cell in the form of visible light as a result of a specific type of chemical reaction within the cell (as opposed to fluorescence or phosphorescence, which will be discussed in the next section). For cells to bioluminescence, they must produce a luciferase enzyme. While many organisms naturally produce this enzyme, typically in cancer research, the small animal is injected with a luciferase-transfected tumor. For small animals injected with luciferase-transfected tumors, if they are then also injected with a luciferin substrate, metabolically active, because the oxygen and ATP are also required for bioluminescence, tumor cells will bioluminescence (emit visible light). Because of the four requirements for bioluminescence (luciferase enzyme, luciferin substrate, oxygen, and ATP), bioluminescence is exquisitely attuned to the presence of tumor cells. For mice that are not themselves transfected with luciferase, only their luciferase-transfected tumor cells will bioluminescence.

The primary advantage of bioluminescence over other preclinical imaging modalities is the extreme specificity of its signal. Since the small animals are placed in a light-tight box, the only light that can be generated in a bioluminescence imaging system is from bioluminescence itself. An additional advantage of bioluminescence is the ability to do low cost, high-throughput imaging. Additionally, bioluminescence can be quantitative for subcutaneous tumors.

The biggest weakness of bioluminescence imaging is that it requires that the targeted cells produce a luciferase enzyme; thus transgenic tumors that express the luciferase enzyme are typically used as part of a research program. Very few studies have directly linked Au NPs and bioluminescence imaging. For the vast majority of studies that involve both bioluminescence imaging and Au NPs, bioluminescence imaging is just used to evaluate the tumoricidal effects of the Au NP-based chemotherapeutic. While this approach is helpful in a research setting, it does not translate into the clinical setting—tumors in cancer patients do not express the luciferase enzyme, so bioluminescence holds no clinical utility. Some groups are working to overcome this weakness through using click chemistry to attach the luciferase enzyme to the Au NP (Kim et al., 2010; Lee & Kim, 2015).

An additional weakness of bioluminescence is that, as it is an optical imaging approach, it suffers from depth effects, even in small animal imaging. While bioluminescence imaging can be quantitative for subcutaneous tumors, most naturally occurring tumors are not subcutaneous. Orthotopic tumor models tend to be semiquantitative, at best, with bioluminescence imaging. Optical light has a very high likelihood of being either scattered or absorbed within tissue, so the vast majority of bioluminescent light generated within deep tissue (such as the liver or lungs) will be absorbed or scattered before it leaves the mouse, and therefore that signal is not measurable.

4.1.2 Fluorescence Imaging

Fluorescence is a process by which molecules called fluorophores absorb and reemit light at characteristic energies. In fluorescence, light energy is absorbed by electrons of the fluorophore, causing the electrons to temporarily move into an excited (higher energy) state. Some energy is initially lost before the electron emits a fluorescence photon, dropping from the excited state back into its rest state. The initial loss of energy is described by the Stokes Shift, and as a result of this energy-loss mechanism, the energy of the fluorescent photon is always less than that of the absorbed photon, which is also to say that the wavelength of the emitted fluorescent light is always longer than the wavelength of the absorbed light. Another key to fluorescence is that the energy of the absorbed light must be sufficient to transition the electron of fluorophore into the excited state; if yellow light is required to put the electron into an excited state, then yellow and higher energy light (green, blue, ultraviolet, etc.) will produce fluorescence, but lower energy light (orange, red, near-infrared, infrared, etc.) will not produce fluorescence.

For in vivo fluorescence imaging of tumors to work, a fluorophore has to localize to the tumor, light of sufficient energy must be able to penetrate deep enough into the tissue to reach the fluorophore, the fluorescent light emitted by the fluorophore must then emerge from the small animal with sufficient brightness to be imaged, and the energy spectrum of the fluorophore must be sufficiently different from the intrinsic energy spectrum of the small animal (energy spectrum of the fluorescence imaging when imaging a control small animal, without the fluorescence probe).

While fluorescence imaging approaches have long been established in microscopy, in vivo fluorescence imaging typically requires unique fluorescence probes. A classic fluorescence probe for microscopy, green fluorescent protein (GFP), works well for thin samples, but green light is strongly scattered or absorbed in tissue, and to make matters worse, collagen (present in the skin and fur of small animals) fluoresces green (Deyl, Praus, Šulcová, & Goldman, 1969), so it can be challenging, or impossible, to image GFP longitudinal in vivo studies. Fluorescence probes that emit near-infrared light are optimal for in vivo small animal imaging—near-infrared light travels much farther without being scattered or absorbed in tissue than light in the optical spectrum (Frangioni, 2003; Haque, Faizi, Rather, & Khan, 2017).

Recently, Xia et al. have shown the effectiveness of fluorescence imaging for measuring the distribution of targeted Au NPs in a tetrafunctional Au NP (Xia et al., 2018). In their study, they used PEG-coated Au NPs that were conjugated with MMP2-modified polypeptides (to target the tumor), with a pH-sensitive doxorubicin, *cis*-aconitic anhydride-modified doxorubicin (to be released in the low-pH vicinity of the tumor), and with chlorin e6 (Ce6; a near-infrared photosensitizer for both fluorescence imaging and photodynamic therapy). Their approach allows for longitudinal assessment of delivery efficiency for the Au NPs as well as assessment of the efficacy of the treatment approach.

4.1.3 Micro-PET

High-resolution positron-emission tomography, called micro-PET, is fundamentally the same imaging approach as clinical PET. In both clinical- and micro-PET, the imaging systems generate 3D maps of the amount of positron-emitting (β+) radioactive isotopes inside of an imaging subject. When positrons (antiparticles of electrons) are spontaneously emitted from a radioactive isotope, they travel some distance (typically ~2–3 mm) before combining with an electron and emitting two 511 keV gamma rays at an angle of 180° apart. The detectors in the PET system then record these two gamma rays and use collections of these gamma ray pairs to map the radioactivity in the subject. Because PET only records these pairs of gamma rays, the background noise of PET systems is very low. Additionally, because gamma rays are only minimally absorbed/scattered in tissue, PET imaging does not suffer nearly as much as optical imaging for deeper tissue imaging. As such, PET is the most sensitive and quantitative of the imaging approaches for in vivo preclinical imaging (Kiessling, Pichler, & Hauff, 2016).

In most PET imaging, the radioactive isotope is inserted into a targeting molecule. By far the most commonly used targeting molecule is fluorodeoxyglucose (FDG) in FDG-PET. FDG is a radiolabeled (with ^{18}F) glucose analog that is treated as glucose by cells upon uptake but is then slowly broken processed by the cells. Because FDG is a glucose analog, FDG-PET is then a measure of metabolic activity within the subject. Regions with the strongest PET signal are the most metabolically active regions; since tumor cells tend to be very metabolically active, FDG-PET can be used to measure the size of a primary tumor and, more importantly, locate metastatic tumors that may be distant to the primary tumor.

While micro-PET is the most quantitative and most sensitive of the preclinical imaging modalities (with subpicomolar sensitivity) (Kiessling et al., 2016), ultimately the cost of running a PET study drastically outpaces the cost of running optical imaging studies (much higher system purchase cost, many studies would require the use of radiochemistry facilities, high cost of some radioisotope, regulatory complexity of working with radioactive isotopes, etc.). As such, optical molecular imaging approaches are far more prevalent in research programs.

4.1.4 Micro-MRI

Dedicated preclinical MRI systems with a high-field strength (7T), which facilitate rapid data acquisition, higher sensitivity, and higher spatial resolution than that of clinical MRI systems, have been shown to be effective at imaging Au NPs in vivo when those Au NPs are conjugated to a contrast-enhancing compound (e.g., gadolinium) (Alric et al., 2008; Shan, 2004–2013).

The advantage of MRI over other preclinical imaging approaches is the nearly depth-independent sensitivity to the gadolinium-enhanced MRI signal (as opposed to the optical imaging approaches, which have rapidly degrading signal as a function of depth), high spatial resolution (much better than PET or SPECT), and improved sensitivity relative to micro-CT.

The disadvantages of MRI relative to optical imaging approaches are a reduced sensitivity, a much higher cost (both in terms of initial purchase price and of long-term maintenance), and a much lower throughput.

4.1.5 Micro-CT

High-resolution computed tomography, called micro-CT, is fundamentally the same imaging approach as clinical CT. In both clinical- and micro-CT, the imaging systems generate 3D maps of the density of the tissues within an imaging subject. In CT, less-dense tissues, such as fat or lung, will appear darker in an image, while more-dense tissue, such as bone, will appear whiter in an image.

Micro-CT can achieve isotropic spatial resolution (i.e., the spatial resolution is the same in all directions) of around 75 μm for in vivo imaging (and much better spatial resolution for ex vivo imaging or for studies in which radiation dose to the small animal is not a concern), while clinical CT typically has about a 250–500 μm in-plane spatial resolution (and ~1–3 mm spatial resolution out of plane).

While tumors embedded in soft tissue tend to be denser than then surrounding soft tissue matrix, often times that difference in density is insufficient to be able to visualize the tumor. In these cases, a contrast agent must be used to visualize the tumor. For any CT contrast agent to be effective, it must selectively enhance the contrast of the feature we would like to image. For imaging of tumors in vivo, enhancing the contrast of the tumor means to increase the density, and thereby the absorption, of the tumor while minimally affecting the absorption of the tissue adjacent to the tumor (remember: for us to see the tumor, it is the *difference in absorption* between the tumor and the adjacent material that matters).

Because gold's density is so high, a Au NP is a good candidate for a contrast agent for micro-CT. The fundamental question is whether or not a sufficient concentration of gold can be selectively delivered to the tumor location. The first use of Au NPs as contrast agents in micro-CT imaging was simply as a blood-pooling contrast agent (colloidal Au NPs; to enhance the contrast of blood vessels) (Cai et al., 2007; Hainfeld, Slatkin, & Smilowitz, 2004). For blood-pooling contrast agent to be effective, it must distribute throughout the vasculature of the body and stay in the blood stream long enough to be imaged. Because small animals have a much more rapid metabolism than humans, blood-pooling contrast agents that work well in humans may not work at all in mice. For example, the most common clinical blood-pooling contrast agent, for example, iohexal (sold under the brand name Omnipaque), is so rapidly cleared by the mouse kidneys that the contrast agent must be continuously perfused into the mouse to be effective (Wathen et al., 2013).

Hainfeld et al. showed that there could be sufficient uptake of Au NPs to visualize the extent of tumors with planar imaging (Hainfeld et al., 2004) and with micro-CT (Hainfeld, Slatkin, Focella, & Smilowitz, 2006). In one study, they showed that untargeted 12 nm Au NPs crossed the blood–brain barrier and were localized to an orthotopic brain tumor through the EPR effect, an effect by which nanoparticles escape from the leaky vasculature of tumors (Hainfeld, Smilowitz, O'Connor, Dilmanian, & Slatkin, 2012). In addition to the enhanced contrast for micro-CT, they also showed a drastic increase in the radiotherapy effectiveness due to the presence of the Au NPs in the tumor (Hainfeld et al., 2012). The same group has coupled anti-Her2 antibodies to 15 nm Au NPs and tested the effectiveness of targeting Her2 + breast cancer cells in a mouse model of breast cancer (Hainfeld et al., 2011). They showed a significant increase in the measured density in CT for the Her2 + tumors over Her2 − tumors, proving the targeting to be effective.

In a very interesting recent study, Si-Mohamed et al. investigated the use of spectral photon-counting computed tomography for the longitudinal assessment of Au NP uptake in rabbits (Si-Mohamed et al., 2017). In phantom-imaging studies, they were able to show excellent sensitivity down to about 1 mg/mL of Au NPs. In the imaging studies of the rabbits, they were able to show enhanced contrast in the rabbit vasculature immediately after injection of the Au NP solution, and uptake of the nanoparticles into the liver, spleen, kidneys, lymph nodes, and bone marrow. Even 6 months postinjection, the Au NPs were still present in the spleen, liver, and bone marrow. While this approach has not yet been specifically tested for cancer, it is a strong candidate for quantifying Au NP uptake and would be directly translatable into the clinic.

4.2 Clinical Imaging

While the focus of preclinical imaging is on high-throughput approaches to assess the efficacy of many different types of treatments (along with many different controls) in the minimum amount of time possible, with clinical imaging, the focus shifts to the sensitivity and specificity of the imaging approach to both localize the primary tumor or metastases and quantify its size. In our considerations below for clinical imaging, we will be speculative, as none of these imaging approaches have yet been validated for the purpose of clinical imaging of Au NPs.

4.2.1 Clinical CT

For quantitative assessment of tumor size or for the assessment of uptake of gold nanoparticle-based chemotherapeutics to be successful, we must be able to deliver approximately 1 mg/mL of gold to the tumor, while only delivering minimal Au NPs elsewhere in the body. But standard clinical CT is unlikely to provide the required specificity for Au NPs—a 1-mg/mL change in local density is observable with CT, but it is an ambiguous result (the change in local density could be due to hardening of the tumor tissue itself). It is likely that any clinical application of CT to quantify the delivery of Au NPs would require the use of either dual-energy CT or spectral CT. Both dual-energy and spectral CT would make it possible to separate the contribution of the Au NPs from the hardened soft tissue and to reduce the ambiguity in the CT data.

4.2.2 Clinical MRI

Au NPs have no intrinsic contrast in MRI, so for clinical MRI to be able to image Au NPs, they need to be tagged with an MRI contrast agent, such as gadolinium. There are several FDA-approved gadolinium formulations (e.g., gadopentetate dimeglumine, sold by Bayer and marketed as Magnevist, and gadodiamide, sold by GE Healthcare and marketed as Omniscan) (Postmarket Drug Safety Information for Patients and Providers, 2018).

It is likely that gadolinium-tagged Au NPs will play an important role in quantifying the efficacy of clinical treatments (Alric et al., 2008; Arifin et al., 2011; Marangoni et al., 2017; Miladi et al., 2014; Shan, 2004–2013). As with clinical CT, clinical MRI suffers from reduced resolution and sensitivity relative to its preclinical counterpart.

4.2.3 Clinical PET

Because of the high sensitivity of PET, it is likely that PET will play a key role in establishing the clinical efficacy of using Au NPs in cancer therapeutics. PET will likely have to play a major role in establishing the pharmacokinetics of the Au NP delivery (Goel, England, Chen, & Cai, 2017). It is likely that a combination of FDG-PET, which is used to find metastatic tumors through their enhanced metabolism, and more tumor-specific imaging, such as targeted, ^{64}Cu-doped Au NPs (Anderson & Ferdani, 2009; Pang et al., 2016), will be used to quantify tumor size and ultimately assess the efficacy of treatment approaches.

5. CONCLUSION

In many ways, the study of Au NPs as delivery vehicles for cancer therapeutics is still in its infancy. The variety of sizes, shapes, charges, coatings, and targeting mechanisms when combined with the nuanced and not entirely understood mechanisms by which the human organ systems process each of these different nanoparticle configurations means that scientists and clinicians have their work cut out for them and then look to develop and evaluate effective Au NP-based chemotherapeutics. While the challenge is immense, the upside of developing targeted chemotherapeutics is far too high to ignore.

REFERENCES

Alric, C., et al. (2008). Gadolinium chelate coated gold nanoparticles as contrast agents for both X-ray computed tomography and magnetic resonance imaging. *Journal of the American Chemical Society*, *130*, 5908–5915.

Anderson, C. J., & Ferdani, R. (2009). Copper-64 radiopharmaceuticals for PET imaging of cancer: advances in preclinical and clinical research. *Cancer Biotherapy & Radiopharmaceuticals*, *24*, 379–393.

Arifin, D. R., et al. (2011). Trimodal gadolinium-gold microcapsules containing pancreatic islet cells restore normoglycemia in diabetic mice and can be tracked by using US, CT, and positive-contrast MR imaging. *Radiology*, *260*, 790–798.

Banu, H., et al. (2015). Doxorubicin loaded polymeric gold nanoparticles targeted to human folate receptor upon laser photothermal therapy potentiates chemotherapy in breast cancer cell lines. *Journal of Photochemistry and Photobiology. B*, *149*, 116–128.

Boisselier, E., & Astruc, D. (2009). Gold nanoparticles in nanomedicine: Preparations, imaging, diagnostics, therapies and toxicity. *Chemical Society Reviews*, *38*, 1759–1782.

Cai, Q.-Y., et al. (2007). Colloidal gold nanoparticles as a blood-pool contrast agent for X-ray computed tomography in mice. *Investigative Radiology*, *42*, 797.

Cai, C., et al. (2018). Transferrin adsorbed on PEGylated gold nanoparticles and its relevance to targeting specificity. *Journal of Nanoscience and Nanotechnology*, *18*, 5306–5313.

Cheng, Y., et al. (2008). Highly efficient drug delivery with gold nanoparticle vectors for in vivo photodynamic therapy of cancer. *Journal of the American Chemical Society*, *130*, 10643–10647.

Cheng, Y., et al. (2010). Delivery and efficacy of a cancer drug as a function of the bond to the gold nanoparticle surface. *Langmuir*, *26*, 2248–2255.

Cheng, Y., et al. (2011). Addressing brain tumors with targeted gold nanoparticles: A new gold standard for hydrophobic drug delivery? *Small (Weinheim an der Bergstrasse, Germany)*, 7, 2301–2306.

Conde, J., et al. (2012). Design of multifunctional gold nanoparticles for in vitro and in vivo gene silencing. *ACS Nano*, *6*, 8316–8324.

Cui, Y., Xu, Q., Chow, P. K.-H., Wang, D., & Wang, C.-H. (2013). Transferrin-conjugated magnetic silica PLGA nanoparticles loaded with doxorubicin and paclitaxel for brain glioma treatment. *Biomaterials*, *34*, 8511–8520.

Demeule, M., et al. (2008). Involvement of the low-density lipoprotein receptor-related protein in the transcytosis of the brain delivery vector angiopep-2. *Journal of Neurochemistry*, *106*, 1534–1544.

Deyl, Z., Praus, R., Šulcová, H., & Goldman, J. N. (1969). Fluorescence of collagen—Properties of tyrosine residues and another fluorescent element in calf skin collagen. *FEBS Letters*, *5*, 187–191.

Dixit, S., et al. (2015). Transferrin receptor-targeted theranostic gold nanoparticles for photosensitizer delivery in brain tumors. *Nanoscale*, 7, 1782–1790.

Frangioni, J. V. (2003). In vivo near-infrared fluorescence imaging. *Current Opinion in Chemical Biology*, 7, 626–634.

Friedman, H. S., Kerby, T., & Calvert, H. (2000). Temozolomide and treatment of malignant glioma. *Clinical Cancer Research: An Official Journal of the American Association for Cancer Research*, *6*, 2585–2597.

Gautier, J., Allard-Vannier, E., Munnier, E., Soucé, M., & Chourpa, I. (2013). Recent advances in theranostic nanocarriers of doxorubicin based on iron oxide and gold nanoparticles. *Journal of Controlled Release: Official Journal of the Controlled Release Society*, *169*, 48–61.

Gibson, J. D., Khanal, B. P., & Zubarev, E. R. (2007). Paclitaxel-functionalized gold nanoparticles. *Journal of the American Chemical Society*, *129*, 11653–11661.

Goel, S., England, C. G., Chen, F., & Cai, W. (2017). Positron emission tomography and nanotechnology: A dynamic duo for cancer theranostics. *Advanced Drug Delivery Reviews*, *113*, 157–176.

Goodman, C. M., McCusker, C. D., Yilmaz, T., & Rotello, V. M. (2004). Toxicity of gold nanoparticles functionalized with cationic and anionic side chains. *Bioconjugate Chemistry*, *15*, 897–900.

Gref, R., et al. (2000). 'Stealth' corona-core nanoparticles surface modified by polyethylene glycol (PEG): Influences of the corona (PEG chain length and surface density) and of the core composition on phagocytic uptake and plasma protein adsorption. *Colloids and Surfaces. B, Biointerfaces*, *18*, 301–313.

Hainfeld, J. F., Slatkin, D. N., Focella, T. M., & Smilowitz, H. M. (2006). Gold nanoparticles: A new X-ray contrast agent. *The British Journal of Radiology*, *79*, 248–253.

Hainfeld, J. F., Slatkin, D. N., & Smilowitz, H. M. (2004). The use of gold nanoparticles to enhance radiotherapy in mice. *Physics in Medicine and Biology*, *49*, N309.

Hainfeld, J. F., Smilowitz, H. M., O'Connor, M. J., Dilmanian, F. A., & Slatkin, D. N. (2012). Gold nanoparticle imaging and radiotherapy of brain tumors in mice. *Nanomedicine (London, England)*, *8*, 1601–1609.

Hainfeld, J. F., et al. (2011). Micro-CT enables microlocalisation and quantification of Her2-targeted gold nanoparticles within tumour regions. *The British Journal of Radiology*, *84*, 526–533.

Haque, A., Faizi, M. S. H., Rather, J. A., & Khan, M. S. (2017). Next generation NIR fluorophores for tumor imaging and fluorescence-guided surgery: A review. *Bioorganic & Medicinal Chemistry*, *25*, 2017–2034.

Harper, S. L., et al. (2011). Systematic evaluation of nanomaterial toxicity: Utility of standardized materials and rapid assays. *ACS Nano*, *5*, 4688–4697.

Haubner, R., et al. (2001). Glycosylated RGD-containing peptides: tracer for tumor targeting and angiogenesis imaging with improved biokinetics. *Journal of Nuclear Medicine*, *42*, 326–336.

Ho, V. K. Y., et al. (2014). Changing incidence and improved survival of gliomas. *European Journal of Cancer*, *50*, 2309–2318.

Hone, D. C., et al. (2002). Generation of cytotoxic singlet oxygen via phthalocyanine-stabilized gold nanoparticles: A potential delivery vehicle for photodynamic therapy. *Langmuir*, *18*, 2985–2987.

Huang, P., et al. (2011). Folic acid-conjugated silica-modified gold nanorods for X-ray/CT imaging-guided dual-mode radiation and photo-thermal therapy. *Biomaterials*, *32*, 9796–9809.

Ito, S., et al. (2009). Enhancement of 5-Aminolevulinic acid-induced oxidative stress on two cancer cell lines by gold nanoparticles. *Free Radical Research*, *43*, 1214–1224.

Jiang, W., et al. (2012). Conjugation of functionalized SPIONs with transferrin for targeting and imaging brain glial tumors in rat model. *PLoS One*, 7, e37376.

Johnston, J. B. et al. Targeting the EGFR pathway for cancer therapy, Current Medicinal Chemistry. (2006), 13, 3483–3492. https://doi.org/10.2174/092986706779026174.

Ju, Y., et al. (2017). Monodisperse au-Fe2C Janus nanoparticles: An attractive multifunctional material for triple-modal imaging-guided tumor photothermal therapy. *ACS Nano*, *11*, 9239–9248.

Karathanasis, E., & Ghaghada, K. B. (2016). Crossing the barrier: Treatment of brain tumors using nanochain particles. *Wiley Interdisciplinary Reviews. Nanomedicine and Nanobiotechnology*, *8*, 678–695.

Kiessling, F., Pichler, B. J., & Hauff, P. (2016). *Small animal imaging*. Springer.

Kim, D., Jeong, Y. Y., & Jon, S. (2010). A drug-loaded aptamer-gold nanoparticle bioconjugate for combined CT imaging and therapy of prostate cancer. *ACS Nano*, *4*, 3689–3696.

Kim, Y.-P., et al. (2010). Bioluminescent nanosensors for protease detection based upon gold nanoparticle-luciferase conjugates. *Chemical Communications (Cambridge, England)*, *46*, 76–78.

Lee, H., & Kim, Y.-P. (2015). Fluorescent and bioluminescent nanoprobes for in vitro and in vivo detection of matrix metalloproteinase activity. *BMB Reports*, *48*, 313–318.

Lewinski, N., Colvin, V., & Drezek, R. (2008). Cytotoxicity of nanoparticles. *Small*, *4*, 26–49.

Maeda, H. (2001). The enhanced permeability and retention (EPR) effect in tumor vasculature: The key role of tumor-selective macromolecular drug targeting. *Advances in Enzyme Regulation*, *41*, 189–207.

Marangoni, V. S., et al. (2017). Enhancing T1 magnetic resonance imaging contrast with internalized gadolinium(III) in a multilayer nanoparticle. *Proceedings of the National Academy of Sciences of the United States of America*, *114*, 6960–6965. https://doi.org/10.1073/pnas.1701944114.

Marko, N. F., et al. (2014). Extent of resection of glioblastoma revisited: Personalized survival modeling facilitates more accurate survival prediction and supports a maximum-safe-resection approach to surgery. *Journal of Clinical Oncology: Official Journal of the American Society of Clinical Oncology*, *32*, 774–782.

Mendes, R., Pedrosa, P., Lima, J. C., Fernandes, A. R., & Baptista, P. V. (2017). Photothermal enhancement of chemotherapy in breast cancer by visible irradiation of gold nanoparticles. *Scientific Reports*, 7(10872), 1–9.

Merali, Z. (2013). This 1,600-year-old goblet shows that the Romans were nanotechnology pioneers. *Smithsonian*. Available at: https://www.smithsonianmag.com/history/this-1600-year-old-goblet-shows-that-the-romans-were-nanotechnology-pioneers-787224/ (Accessed: 15th March 2018).

Miladi, I., et al. (2014). The in vivo radiosensitizing effect of gold nanoparticles based MRI contrast agents. *Small (Weinheim an der Bergstrasse, Germany)*, *10*, 1116–1124.

Murawala, P., Tirmale, A., Shiras, A., & Prasad, B. L. V. (2014). In situ synthesized BSA capped gold nanoparticles: Effective carrier of anticancer drug methotrexate to MCF-7 breast cancer cells. *Materials Science & Engineering. C, Materials for Biological Applications*, *34*, 158–167.

Murphy, C. J., et al. (2008). Gold nanoparticles in biology: Beyond toxicity to cellular imaging. *Accounts of Chemical Research*, *41*, 1721–1730.

Nava, F., et al. (2014). Survival effect of first- and second-line treatments for patients with primary glioblastoma: A cohort study from a prospective registry, 1997-2010. *Neuro-oncology*, *16*, 719–727.

Niu, W., Chua, Y. A. A., Zhang, W., Huang, H., & Lu, X. (2015). Highly symmetric gold nanostars: Crystallographic control and surface-enhanced Raman scattering property. *Journal of the American Chemical Society*, *137*, 10460–10463.

Noone, A. M., Howlader, N., Krapcho, M., Miller, D., Brest, A., Yu, M., Ruhl, J., Tatalovich, Z., Mariotto, A., Lewis, D. R., Chen, H. S., Feuer, E. J., & Cronin, K. A. (Eds.), (1975–2015). *SEER Cancer Statistics Review*. Bethesda, MD: National Cancer Institute. https://seer.cancer.gov/csr/1975_2015/. Based on November 2017 SEER data submission, posted to the SEER web site, April 2018.

Oo, M. K. K., Yang, X., Du, H., & Wang, H. (2008). 5-Aminolevulinic acid-conjugated gold nanoparticles for photodynamic therapy of cancer. *Nanomedicine (London, England)*, *3*, 777–786.

Pang, B., et al. (2016). ^{64}Cu-doped PdCu@Au tripods: A multifunctional nanomaterial for positron emission tomography and image-guided photothermal cancer treatment. *ACS Nano*, *10*, 3121–3131.

Peralta, D. V., Heidari, Z., Dash, S., & Tarr, M. A. (2015). Hybrid paclitaxel and gold nanorod-loaded human serum albumin nanoparticles for simultaneous chemotherapeutic and photothermal therapy on 4T1 breast cancer cells. *ACS Applied Materials & Interfaces*, 7, 7101–7111.

Postmarket Drug Safety Information for Patients and Providers. (2018). Postmarket drug safety information for patients and providers—Information on gadolinium-based contrast agents. Available at: https://www.fda.gov/Drugs/DrugSafety/PostmarketDrugSafetyInformationforPatientsandProviders/ucm142882.htm. Federal Drug Administration, (Accessed: 17th March 2018).

Schaeublin, N. M., et al. (2011). Surface charge of gold nanoparticles mediates mechanism of toxicity. *Nanoscale*, *3*, 410–420.

Schuemann, J., et al. (2016). Roadmap to clinical use of gold nanoparticles for radiosensitization. *International Journal of Radiation Oncology, Biology, Physics*, *94*, 189–205.

Shan, L. (2004–2013). Gold nanoparticles coated with dithiolated diethylenetriamine pentaacetic acid-gadolinium chelate. 2010 Sep 24 [Updated 2010 Nov 17]. In: *Molecular Imaging and Contrast Agent Database (MICAD) [Internet]*. Bethesda (MD): National Center for Biotechnology Information (US). Available from: https://www.ncbi.nlm.nih.gov/books/NBK49244/.

Si-Mohamed, S., et al. (2017). Evaluation of spectral photon counting computed tomography K-edge imaging for determination of gold nanoparticle biodistribution in vivo. *Nanoscale*, *9*, 18246–18257.

Stupp, R., et al. (2005). Radiotherapy plus concomitant and adjuvant temozolomide for glioblastoma. *The New England Journal of Medicine*, *352*, 987–996.

Tkachenko, A. G., et al. (2004). Cellular trajectories of peptide-modified gold particle complexes: Comparison of nuclear localization signals and peptide transduction domains. *Bioconjugate Chemistry*, *15*, 482–490.

Torchilin, V. P. (2006). Multifunctional nanocarriers. *Advanced Drug Delivery Reviews*, *58*, 1532–1555.

Wang, C., Jiang, Y., Li, X., & Hu, L. (2015). Thioglucose-bound gold nanoparticles increase the radiosensitivity of a triple-negative breast cancer cell line (MDA-MB-231). *Breast Cancer*, *22*, 413–420.

Wang, D., et al. (2014). Treatment of metastatic breast cancer by combination of chemotherapy and photothermal ablation using doxorubicin-loaded DNA wrapped gold nanorods. *Biomaterials*, *35*, 8374–8384.

Wathen, C. A., et al. (2013). In vivo X-ray computed tomographic imaging of soft tissue with native, intravenous, or oral contrast. *Sensors*, *13*, 6957–6980.

Wolfe, T., et al. (2015). Targeted gold nanoparticles enhance sensitization of prostate tumors to megavoltage radiation therapy in vivo. *Nanomedicine: Nanotechnology, Biology, and Medicine*, *11*, 1277–1283.

Wu, P. H., et al. (2017). Targeting integrins with RGD-conjugated gold nanoparticles in radiotherapy decreases the invasive activity of breast cancer cells. *International Journal of Nanomedicine*, *12*(12), 5069–5085.

Xia, F., et al. (2018). pH-responsive gold nanoclusters-based nanoprobes for lung cancer targeted near-infrared fluorescence imaging and chemo-photodynamic therapy. *Acta Biomaterialia*, *68*, 308–319.

Xiong, H. Q., et al. (2004). Cetuximab, a monoclonal antibody targeting the epidermal growth factor receptor, in combination with gemcitabine for advanced pancreatic cancer: A multicenter phase II trial. *Journal of Clinical Oncology*, *22*, 2610–2616.

Yan, J.-J., Liao, J.-Z., Lin, J.-S., & He, X.-X. (2015). Active radar guides missile to its target: Receptor-based targeted treatment of hepatocellular carcinoma by nanoparticulate systems. *Tumor Biology*, *36*, 55–67.

Yu, F., et al. (2016). Glutathione-responsive multilayer coated gold nanoparticles for targeted gene delivery. *Journal of Biomedical Nanotechnology*, *12*, 503–515.

Zhang, Y.-N., Poon, W., Tavares, A. J., McGilvray, I. D., & Chan, W. C. W. (2016). Nanoparticle-liver interactions: Cellular uptake and hepatobiliary elimination. *Journal of Controlled Release: Official Journal of the Controlled Release Society*, *240*, 332–348.

Zhang, H., et al. (2018). Gold nanorods conjugated porous silicon nanoparticles encapsulated in calcium alginate Nano hydrogels using microemulsion templates. *Nano Letters*, *18*, 1448–1453.

CHAPTER EIGHT

Single-Cell Proteomics for Cancer Immunotherapy

Linmei Li*, Shiqiang Yan*, Bingcheng Lin*, Qihui Shi†,‡,1, Yao Lu*,1

*Dalian Institute of Chemical Physics, Chinese Academy of Sciences, Dalian, China
†Minhang Branch, Zhongshan Hospital and Institute of Biomedical Sciences, Fudan University, Shanghai, China
‡Institute of Fudan-Minhang Academic Health System, Minhang Hospital, Fudan University, Shanghai, China
[1]Corresponding authors: e-mail address: qihuishi@fudan.edu.cn; luyao@dicp.ac.cn

Contents

Abstract

Cancer immunotherapy fights against cancer by modulating the immune response and is delivering encouraging results in clinical treatments. However, it is challenging to achieve durable response in all cancer patients during treatment due to the diversity and dynamic nature of immune system as well as inter- and intratumor heterogeneity. A comprehensive assessment of system immunity and tumor microenvironment is crucial for effective and safe cancer therapy, which can potentially be resolved by single-cell proteomic analysis. Single-cell proteomic technologies enable system-wide profiling of protein levels in a number of single cells within the immune system and tumor microenvironment, and thereby provide direct assessment of the functional state of the immune cells and tumor–immune interaction that could be used to evaluate efficacy of immunotherapy and to improve clinical outcome. In this chapter, we summarized current single-cell proteomic technologies and their applications in cancer immunotherapy.

Advances in Cancer Research, Volume 139
ISSN 0065-230X
https://doi.org/10.1016/bs.acr.2018.04.006

1. INTRODUCTION

Cancer immunotherapy is a promising new way to treat cancer by modulating the immune response (Smith & Cerundolo, 2001). Immunotherapy strategies, including immune-cell-targeted monoclonal antibody (mAb) therapy and adoptive cell transfer (ACT) therapy, have led to unprecedented responses in patients with advanced-stage tumors (Dillman, 2011; Restifo, Dudley, & Rosenberg, 2012; Rosenberg, Restifo, Yang, Morgan, & Dudley, 2008). These successes suggest that immunotherapy represents a path to obtain a durable and long-lasting response (potentially a cure) in cancer patients.

The immune-cell-targeted mAb therapy is based on stimulation of T-cell function by blocking or activating immune regulatory pathways. In contract with traditional cancer therapies, immune modulatory antibodies target immune cells rather than cancer cells, for example, immune checkpoint blockade, a method to block immune inhibitory pathways with antibodies. The goal of immune checkpoint therapy is to remove inhibitory pathways that block effective antitumor T cell responses rather than to activate the immune system for attacking cancer cells. To date, several immune checkpoint-blocking mAbs have been approved by the U.S. Food and Drug Administration (FDA) for cancer treatment. For example, ipilimumab, a blockage of CTLA-4, was found to improve the overall survival of metastatic melanoma patients and approved by FDA in 2011. Antiprogrammed death-1 (PD-1) immunotherapy that blocks the signaling between PD-1 and its ligands (PD-L1 as one major ligand) has achieved significant clinical responses in multiple tumor types (Topalian et al., 2012), leading to approval of two antibodies against PD-1 (pembrolizumab and nivolumab) in 2014. Although these inspiring clinical successes have been made, durable clinical responses have been shown only in a fraction of patients. It is thereby of immense importance to identify predictive biomarkers for selection of patients who would benefit from the therapy. However, the complexity of tumor microenvironment and the dynamic nature of the immune response make this attempt difficult and require a new tool to resolve the immune response and tumor–immune interaction at the single-cell level.

The ACT therapy represents a different therapeutic strategy that aims to generate a robust antitumor response through genetically engineered T cells. These ex vivo manipulated T cells are capable of recognizing cancer cells by genetically engineered cancer-specific, synthetic T-cell receptor (TCR), or chimeric antigen receptor (CAR). Kymriah (tisagenlecleucel), the first chimeric antigen receptor T cell (CAR-T) therapy, was approved by FDA in 2017 for the treatment of patients with B-cell precursor acute lymphoblastic leukemia that was refractory or in the second or later relapse. Similar with other cancer therapies, only a fraction of patients have responses to ACT therapy because of not only the heterogeneity in tumors but also the functional heterogeneity of ex vivo manipulated T cells (Marusyk, Almendro, & Polyak, 2012; Porter et al., 2015). For example, the efficiency of ACT therapy is highly dependent on the antitumor activity of CAR-T or TCR-T cells (Johnson et al., 2009; Morgan et al., 2006), and such antitumor activity is found to be correlated to the frequency and levels of antitumor effector proteins (e.g., granzyme B, TNF-α, and IFN-γ) of genetically engineered T cells after their infusion into melanoma patients (Ma et al., 2013). Profiling antitumor effector proteins requires a single-cell proteomic tool to characterize the functional states and heterogeneity of those genetically engineered, therapeutic cells for improving clinical benefits of ACT therapy.

Single-cell proteomics represents the most common and effective method to investigate cellular function and heterogeneity by system-wide assaying functional protein levels in a number of single cells from different cell types. Single-cell proteomics provides a clinical tool for assessment of immune system, ex vivo manipulated immune cells, and tumor microenvironment. In this chapter, we highlight the latest advances in single-cell proteomic technologies and their clinical applications in cancer immunotherapy with the discussion of the role immunotherapy will have in the future.

2. SINGLE-CELL PROTEOMIC ANALYSIS TECHNOLOGIES

Proteins, which are the direct players of various cellular functions, are typically generated or released in multiple, transient pulses, and low abundance (10^2–10^4 copies/cell). Due to the lack of efficient amplification methods comparable as PCR (polymerase chain reaction) for genomic analysis, sensitive, multiplexed profiling technology is highly desired for single-cell proteomic analysis, which will be of great help to address many important biological and medical questions ranging from immune diversity, intratumor heterogeneity, to cell–cell communication. Currently, most commonly used techniques for single-cell proteomic analysis include flow cytometry, enzyme-linked immunospot (ELISpot) and microchip-based technologies, like microengraving and single-cell barcode microchip (SCBC), which be summarized in the following section.

2.1 Flow Cytometry Technologies

2.1.1 Fluorescence Flow Cytometry

Fluorescence flow cytometry (FFC) or fluorescence-activated cell sorting (FACS), dated back to 1970s, is one of the most powerful tools for single-cell phenotyping and functional proteins analysis, which is based on labeling of target proteins in cells with fluorophore-conjugated antibodies (Table 1) to be detected one by one at high-throughput ($\geq$10,000 cells/s) by laser-activated fluorescence. Right now, around 10–17 proteins (mostly for surface markers, 3–5 intracellular proteins) can be simultaneously detected at single-cell level (De Rosa, Herzenberg, Herzenberg, & Roederer, 2001; Perfetto, Chattopadhyay, & Roederer, 2004). The degree of multiplexing can be further increased with the development of new fluorescence dyes and staining methods (Herzenberg et al., 2002; Krutzik & Nolan, 2006). However, limited by severe spectral overlap with increased fluorescence dyes, at most 17 proteomic parameters (with additional 2 physical parameters) can be analyzed simultaneously for each single cell (Perfetto et al., 2004).

FACS is widely used for a lot of applications, like cell counting, phenotyping, and sorting. For example, immune cells phenotyping, which can be correlated with different cell surface markers (Betts et al., 2006; De Rosa et al., 2001; Irish et al., 2010), has been used for characterization of circulating T cells in melanoma patients to discriminate the heterogeneity between primary and metastatic prostate cancer by recognizing different cluster designation (CD) antigens (Bendall & Nolan, 2012; Liu, Roudier, & True, 2004). However, surface markers cannot uncover intracellular signaling events responding to stimulus. Fortunately, intracellular functional proteins information can also be obtained using intracellular cytokine-staining method coupled with FFC detection, in which multiple signaling proteins like cytokines, phosphoproteins can be analyzed concurrently with surface phenotyping. And it has been widely employed in clinical settings, for instance, analysis of the capacity of the immune system response to disease (Betts et al., 2006; Irish et al., 2010; Krutzik, Irish, Nolan, & Perez, 2004), tumor diagnosis (Bhagwat & Carpenter, 2017; Cordone et al., 2016; Urbanits et al., 2002), and tumor progression (Irish et al., 2010; Verronese et al., 2016).

Table 1 Single-Cell Proteomic Analysis Technologies

Technique	Multiplexity for Proteins (Cytokines)	Throughput	Detected Traits	Readout	Cell Recovery	Reference
Flow cytometry techniques						
Fluorescence flow cytometry	Around 15 proteins (3–5 intracellular cytokines)	10^4 cells/s	Surface markers and intracellular cytokine or phosphoprotein	Fluorescence (antibody staining)	Yes	Herzenberg et al. (2002) and Bendall, Nolan, Roederer, and Chattopadhyay (2012)
Mass flow cytometry	40+ Proteins (~10 intracellular cytokines)	10^3 cells/s	Surface markers and intracellular cytokines or phosphoproteins	Isotope (rare metal labeling)	No	Spitzer and Nolan (2016)
ELISpot	1–3 Secreted cytokines	10^6–10^7 cells per dish	Secreted cytokines	Enzyme, fluorescence (antibody labeling)	No	Clifton et al. (2017) and Obara et al. (2017)
Microchip-based techniques						
Image cytometry	3–5 Surface marker or intracellular proteins	10^3–10^4 cells per chip	Surface markers and intracellular cytokine	Fluorescence (antibody staining)	No	Sun et al. (2010) and Zrazhevskiy and Gao (2013)

Continued

Table 1 Single-Cell Proteomic Analysis Technologies—cont'd

Technique	Multiplexity for Proteins (Cytokines)	Throughput	Detected Traits	Readout	Cell Recovery	Reference
Microengraving	3 Secreted cytokines	10^3–10^5 cells per chip	Surface markers, antibodies, and secreted cytokine	Fluorescence (antibody labeling)	Yes	Love, Ronan, Grotenbreg, van der Veen, and Ploegh (2006) and Torres, Contento, Gordo, Wucherpfennig, and Love (2013)
Single-cell barcode chip	42 Secreted cytokines	10^3–10^5 cells per chip	Secreted cytokine and intracellular phosphoproteins	Fluorescence (antibody labeling)	Yes	Fan et al. (2008) and Lu et al. (2015)
Single-cell western blotting	10+ Surface markers and proteins (isoform)	10^3–10^4 cells per chip	Surface markers and intracellular proteins	Fluorescence	No	Hughes et al. (2014) and Tentori, Yamauchi, and Herr (2016)
DNA barcoding method						
DNA barcoded antibody	90+ Surface markers and intracellular proteins		Surface markers and intracellular proteins	DNA quantity or fluorescence	No	Agasti, Liong, Peterson, Lee, and Weissleder (2012) and Ullal et al. (2014)

2.1.2 Mass Cytometry

Mass cytometry, also called cytometry by time-of-flight, is an emerging powerful single-cell proteomic analysis technique which utilizes rare metal isotopes instead of fluorophores for antibodies labeling to break the limit of multiplexing capability of FACS (Table 1; Fig. 1A) (Bandura et al., 2009; Bendall et al., 2011). Due to the minimal overlap between metal isotopes, 40+ protein parameters can be simultaneously quantified within each single cell to investigate human hematopoietic single cells in response to different drugs (Bendall et al., 2011). However, the throughput (~1000 cells/s) and sensitivity of mass cytometry are still lower than conventional FACS. Besides, as the cells are totally "evaporated" during the assay, the cells analyzed with mass cytometry cannot be retrieved for downstream analysis (Spitzer & Nolan, 2016).

The more abundant proteomic information from each single cell enabled by mass cytometry provides a much broader landscape for different types of biological researches, such as the frequency of immune cells response to a stimulus (Fisher et al., 2017), the correlation between tumor pathology and phosphoproteins signaling (Spitzer & Nolan, 2016). Some significant clinical information can also be obtained, including evaluating immune response to disease (Kaiser et al., 2017; O'Gorman et al., 2017), assessment of clinical recovery, and effective disease therapy guidance (Gaudilliere et al., 2014; Nair et al., 2015).

2.2 Enzyme-Linked Immunospot

The ELISpot assay is a widely used, sensitive method for single-cell cytokine secretion monitoring based on sandwich immunoassay. Briefly, cells are added into a membrane-backed microplate (mostly PVDF membrane or nitrocellulose membrane), which is precoated with primary antibodies of target proteins (1–3 cytokines) (Table 1; Fig. 1B). After incubated for certain period of time to allow secreted proteins be captured by the antibodies precoated, cells are removed and the protein signals would be converted into detectable signals by sandwich-type immunoassay and staining to get the "footprint" of secreted cells, in which each spot is indicative of a single cell with positive secretion and the frequency of secretion can be quantified. The total number of cells is an important factor to ensure analysis sensitivity, typically, 200,000–400,000 cells are needed (Britten et al., 2008). However, only 1–3 cytokines can be codetected limited by spectral overlap.

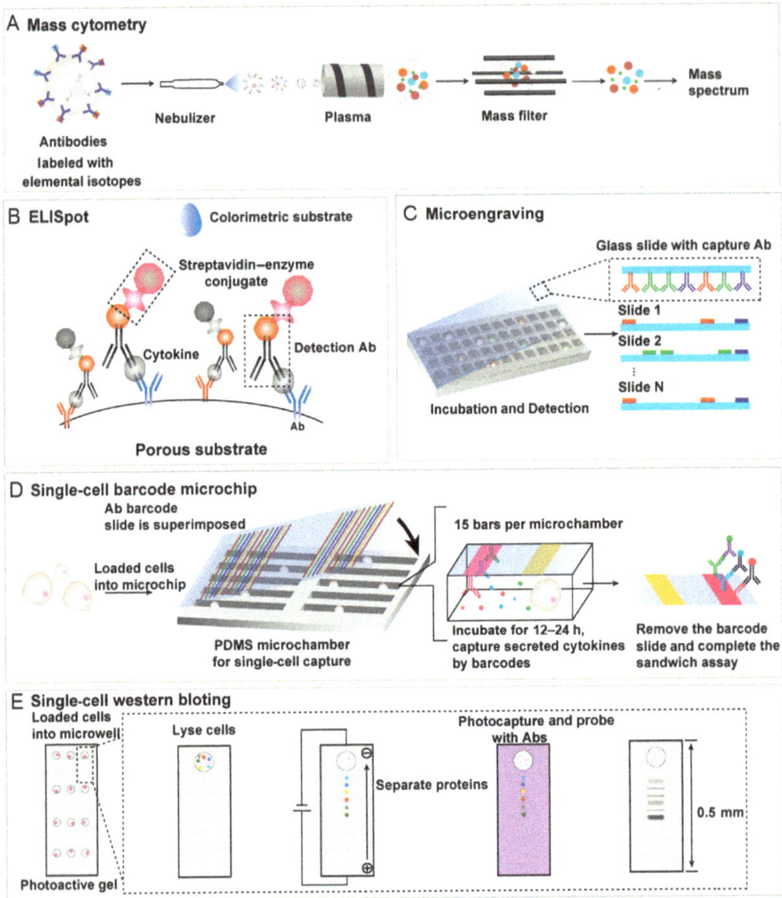

Fig. 1 Representative single-cell proteomic analysis technologies. (A) Mass cytometry. Cells are stained with antibodies conjugated to metal isotope reporters with different mass. Then cells divided into single cell droplets are nebulized and analyzed by mass spectrometry. (B) ELISpot. Cells are incubated in antibody precoated porous well plate and the resulting cytokine secretion will be captured and transformed into positive staining spots based on sandwich immunoassay. (C) Microengraving. Single cells are distributed into microwells, and their protein secretion are recorded with antibody-coated glass slide. Dynamic detection of cytokine secretion at different time points can be achieved via repeating microengraving process with new antibody-coated glass slides. (D) Single-cell barcode microchip. Single cells are isolated into high-throughput microchambers covered with antibody barcode glass slide and cultured for 12–24h. 42 Secreted proteins can be simultaneously profiled for each single cell by combing spatial and spectral encoding. (E) Single-cell western blotting. Single cells are captured into polyacrylamide (PMA) gel microwells array and go through in situ cell lysis, gel electrophoresis for protein separation, UV-induced protein immobilization sequentially. Intracellular proteins of each cell are detected via fluorophore conjugated antibodies and quantified by fluorescence readout.

ELISpot has been widely adopted for monitoring rare antigen-specific T cells and B cells in responses to vaccine in cancer patient (Lewis et al., 2000; Obara et al., 2017; Smith & Cerundolo, 2001) and assessment of immunologic responses to trastuzumab (a mAb) treatment in breast-cancer patients via IFN-γ ELISpot (Clifton et al., 2017).

2.3 Microchip-Based Technologies

2.3.1 Image Cytometry

Image cytometry, also called microfluidic image cytometry, is a technique which combines the power of imaging analysis and the advantage of microfluidics. Microfluidic tools provide distinct advantages of precise control and manipulation of fluids in a submicroliter volume, which makes the sample consumption down to 1000–2800 cells for each assay (Table 1). Such low sample/reagent consumption, cost efficiency, as well as high-throughput makes image cytometry as a perfect fit for rare sample clinical analysis. Sun et al. presented a polydimethylsiloxane (PDMS) microfluidic chip designed with 24 microchannels array for image cytometry, in which different steps of immunostaining like sample loading, immunolabeling, fluorescence staining, and cell imaging can be performed on chip. Four functional proteins (EGFR, PTEN, phospho-Akt, and phospho-S6) and their related single-cell signaling pathway were studied with 19 human brain tumor specimens by immunocytochemistry, which predicted tumor progression and patient survival (Sun et al., 2010). Time-resolved analysis of cellular response under a large number of genetic perturbations and stimulus can also be obtained with microfluidic imaging platform (Taylor et al., 2009). Fan group improved both sensitivity and dynamic range of image cytometry using laser scanner as an alternative detector with high sensitivity (Zhou, Wu, Lee, & Fan, 2012). To further increase the number of parameters codetected, Zrazhevskiy et al. developed a highly multiplexing imaging technology with repeated staining/destaining cycles, in which fluorescent signals from quantum dot-linked antibodies can be destaining via low-H/detergent-based regeneration buffer. Up to 20 intracellular proteins from each single cell can be profiled in a five-color and four-cycle protocol and more multiplicity can be achieved via more staining/destaining cycles (Zrazhevskiy & Gao, 2013).

2.3.2 Microengraving Method

Microengraving, based on isolating single cells in microwells (0.1–1 nL) with a glass slide coated with antigen or primary antibody (Table 1; Fig. 1C), is a miniaturized, high-throughput FluoroSpot assay (Bradshaw et al., 2008; Love et al., 2006), from which the information including surface marker, cytokines, and antibodies can be simultaneously obtained from the same single cell. Due to the concentration of proteins is significantly enriched in the enclosed small volume, low abundance proteins (lower than 10^3 copies/cell) in single cell can be successfully detected. The number of microwells (>100,000 individual cells) ensures high-throughput measurements in a single run. A unique advantage with this method is that dynamic analysis or sequential detection of multiparameters (such as cytokine and antibody) can be easily achieved via repeating microengraving process with new antibody-coated glass slides (Bradshaw et al., 2008; Han et al., 2012). For example, Love's groups applied this method to investigate the immune cells secretion dynamics (Han et al., 2012) and immune cells functional response to diseases or artificial ligands and predicting disease progress (Torres et al., 2013; Varadarajan et al., 2011). When combined with fluorescent cellular barcodes, the throughput of single-cell secretory measurements can be significantly increased (Yamanaka et al., 2012). Just as spectral overlap limit encountered in ELISpot, only four secreted proteins can be detected simultaneously.

The single cells retained in microwells can also be collected for further analysis. Shi group developed a microchip containing 200,000 addressable microwells for rapid and accurate rare tumor diagnosis, which identified rare metabolically active tumor cells in pleural effusion and peripheral blood of lung cancer patients. The sample cells were labeled with CD45/2-NBDG (a fluorescent glucose analog)/EthD-1 (fluorescent dead cell marker) and the rare metabolically active tumor cells in the sample can be distinguished via screening CD45 surface marker and enhanced glucose uptake, a hallmark of cancer cells. Abnormal tumor cells in the microwells can be retrieved for downstream genetic analysis, which provides a powerful tool for clinical diagnosis of malignant pleural effusion that is associated with metastatic malignancies.

2.3.3 Single-Cell Barcode Microchip

SCBC, developed by Heath group, is a highly multiplexed proteomic analysis technology enabled by spatially immobilized primary antibodies barcode with high uniformity and density, which are patterned onto poly-L-lysine (PLL) glass slide via PDMS microchip with parallel microchannels array. Single cells are distributed into corresponding elongated microchambers array in nanoliter volume and enclosed with antibodies barcode glass slide

(Table 1; Fig. 1D). Proteins from the cells would be captured by antibodies and read out by sandwich immunoassay. The numbers of protein molecules detected can be quantified according to the fluorescence levels using respective calibration curves. The antibody immobilization was first accomplished with DNA-encoded antibody libraries (DEAL), which is combined with patterning complemental sDNA oligomers onto the PLL glass slide, and used for detection of cells, DNAs, and proteins (Bailey, Kwong, Radu, Witte, & Heath, 2007). Ma et al. developed the first generation of SCBC with DEAL method for ex vivo quantification of T cell polyfunctional diversity via the simultaneous measurement of a dozen protein molecules secreted from tumor antigen-specific cytotoxic T lymphocytes (CTLs) (Ma et al., 2011). Later on, Lu et al. reported 14-plexed protein detection with direct antibody patterning to investigate the correlation between secretomic signature and the cells' physical behavior (migratory) (Lu et al., 2013). By combining the advantage of spatial encoding (15 barcode array) and spectral encoding (3 colors) (Table 1), they further increased protein parameters to 42 from each single immune cell, representing the highest multiplexing capability record to date for live single-cell secretion detection (Lu et al., 2015).

The SCBC cannot only be applied for the characterization of cells phenotypes subsets and secreted cytokines but also for the cytoplasmic phosphoprotein signaling network research. Shi et al. developed a SCBC platform with pneumatic valve layer integrated for flow control (Shi et al., 2012; Shin et al., 2010), in which cell loading, lysis, as well as cytoplasmic analysis are all realized on this integrated microchip. They quantified 11 intracellular proteins to uncover EGF-receptor-mediated PI3K signaling pathway in single cancer cell (Shi et al., 2012). Xue et al. developed surface-competitive-binding assays on SCBC chip for intracellular metabolite quantification, such as glucose uptake and proteins (Xue, Wei, et al., 2015), from which they revealed both the intercellular metabolic heterogeneity and metabolite–phosphoprotein correlation network of glioblastoma cancer cells in response to drug.

Besides single cell data, small number of cells can also be isolated in microchambers on the SCBC platform, which provides new venue for cell–cell communication research (such as autocrine and paracrine signaling). Xue et al. explored paracrine signaling in macrophage responses to TLR4 stimulation with SCBC platform to reveal a significant role for cell-to-cell communication within a population in coordinating a rapid innate immune response (Xue, Lu, et al., 2015); Elitas et al. explored paracrine signaling between macrophage and glioma cells to reveal the key signaling nodes altered by tumor–macrophage communication (Elitas, Brower, Lu, Chen, & Fan, 2014).

2.3.4 Single-Cell Western Blotting

Single-cell western blotting (scWestern), developed by Herr group in 2014, is a miniaturized analysis tool for single-cell intracellular proteins based on conventional western blotting and immunoassay (Hughes et al., 2014). Single cells, captured into polyacrylamide (PMA) gel microwells array (6720 microwells), went through in situ cell lysis, gel electrophoresis, UV-active protein immobilization, and in-gel immunoprobing sequentially (Table 1; Fig. 1E). The proteins in cell lysates were first separated by molecular mass (via electrophoresis) prior to immunoprobing to discriminate target proteins from off-target signals and detected via fluorophore conjugated antibody and quantified by fluorescence readout with limit of detection down to ~27,000 molecules. Multiplexed (11 proteins) protein detection can be achieved in the same cell lysate via stripping and reprobing rounds. scWestern has been applied in biological research and clinical assessment, including monitoring single rat neural stem cell differentiation (Hughes et al., 2014), assessing glioblastoma heterogeneity response to chemotherapy (Kang, Lin, Xu, Kumar, & Herr, 2014), as well as multiplexed protein analysis for circulating tumor cells (Sinkala et al., 2017).

2.4 DNA Barcoding Method

DNA barcoding-based protein measurement method transforms the protein signals into barcoded oligo probes and amplifies the signals by nucleic acid amplification to achieve ultrasensitive detection, even down to single-molecule level (Konry, Smolina, Yarmush, Irimia, & Yarmush, 2011), which is ideal for detection of low-abundance surface markers of cancer cell or rare tumor cells (Kazane et al., 2012; Konry et al., 2011). Weissleder group developed photocleavable DNA-barcoded antibodies to specifically recognizing multiplexed biomarker of cells. DNA can be cleaved by light (~365 nm) and released into solution to be amplified and analyzed using gel electrophoresis (Agasti et al., 2012). They further developed amplification-free method to profile 90+ proteins in single cells using DNA-barcoded antibodies with fluorescent readout by NanoString to investigate inter- and intratumor heterogeneity as well as the pathway to drug response in clinical samples (Ullal et al., 2014; Ullal & Weissleder, 2015).

3. CLINICAL APPLICATIONS IN CANCER IMMUNOTHERAPY

Cancer immunotherapy is a rapidly advancing cancer treatment method that exploits human immune system to fight against cancer. After many years of continuous efforts, immunotherapy has developed from non-specific therapy (cytokines) to targeted therapy including immune checkpoint blockade and ACT (Dillman, 2011) and now becomes a revolutionary treatment for several cancers. Although inspiring clinical achievements have been made, patient response toward immunotherapy varies significantly and only a fraction of patients have response to this therapy (Hodi et al., 2010) because of the dynamic nature of immune cells and complex tumor–immune interaction within the tumor microenvironment. Single-cell proteomic method enables characterization of functional states of immune cells, especially genetically engineered immune cells, with single-cell resolution and the tumor–immune interaction within the tumor microenvironment, and thereby allows to identify responders who can benefit from the immunotherapy and improve current immunotherapies (Table 2).

3.1 Immune Checkpoint Blockade

Immune checkpoints are members of cell surface receptors which regulate the immune system, including stimulatory and inhibitory checkpoints. Inhibitory checkpoints play a crucial role in downregulation of immune system, and high expression of inhibitory checkpoints in T cells suppresses antitumor CTL activity (Simpson et al., 2013). Thus, immune checkpoint blockade has been increasingly considered as new strategy for cancer immunotherapy. In respect of the clinical success to date, two immune checkpoint blockades, anti-CTLA-4 and anti-PD-1, are clinically used.

Krieg et al. used the high-dimensional, single-cell mass cytometry to characterize subsets of immune cells in the peripheral blood samples taken from metastatic melanoma patients before and 12-week after anti-PD-1 immunotherapy. Thirty surface markers were first assayed in the leukocytes and a set of T cell subsets at the different stages of differentiation and activation. Then, functional phenotypes of T cells and myeloid cells were extensively characterized with single-cell proteomic profiling to discover the difference between responders and nonresponders of the anti-PD-1 immunotherapy. It was found that the frequency of $CD14^{+}$ $CD16^{-}$ $HLA\text{-}DR^{hi}$ monocytes could potentially be used as a prognostic biomarker

Table 2 Single-Cell Proteomics for Cancer Immunotherapy Applications

Assay	Sample	Analysis Targets	Measurement Method	Results	Potential Application	Reference
Immune checkpoint blockade immunotherapy						
Anti-PD-1	PBMCs from healthy donors and human melanoma before and 12 weeks after anti-PD-1 therapy	30 Surface markers and 8 proteins	Mass cytometry for phenotyping and intracellular proteins analysis of T cells	Predict biomarker for anti-PD-1 therapy outcome by the frequency of $CD14^{+}$ $CD16^{-}$ HLA-DR^{hi} monocytes in PBMCs	Discriminate responders from nonresponders before anti-PD-1 therapy	Krieg et al. (2018)
Anti-PD-1 or anti-CTLA-4	Melanoma tumors from human and murine tumor models	33 Surface markers for melanoma tumor in human and mouse	Mass cytometry for phenotyping and proteins analysis of T cells	Anti-PD-1 blockade expanding exhausted-like CD8 T cells, anti-CTLA-4 expanding an $ICOS^{+}$ Th1-like CD4 subset and exhausted like CD8 T cells	Two blockades showing distinct cell mechanisms to guiding the effective therapy	Wei et al. (2017)
Anti-PD-1 (control) or tumor-binding allogeneic IgG	Lymph node, spleen, blood, and bone marrow from murine breast cancer model	39 Parameters for surface markers and proteins	Mass cytometry for cells phenotype subsets and proteins detection in samples	Effective cancer immunotherapy requires immune activation in the periphery	Systemic assessment immune responses and clinical outcome	Spitzer et al. (2017)
ACT immunotherapy						
MART-1 TCR-T	PBMCs from health donors and TCR-T therapy of human metastatic melanoma	8+ Surface markers and 12+ secreted proteins of T cells	FFC for T cells phenotyping and SCBC for T cells secreted proteins detection	High-functional heterogeneity and changes of TCR-T cells (number, phenotype, and antitumor effector)	Assessment antitumor immunity efficacy of TCR-T therapy	Ma et al. (2013, 2011)

CD19 CAR-T	Genetically engineered CD19 CAR-T cells from healthy PBMCs	4 Surface markers and 16 proteins of T cells	Image cytometry for CAR-T subsets, SCBC for their secreted proteins detection	Heterogeneity of polyfunctional subsets in phenotypically similar $CD19^+$ CAR-T cells from donors	Comprehensive assessment antitumor immunity for efficacy or safety CAR-T cell therapy	Xue et al. (2017)
Tumor heterogeneity						
Lymphoma (FL) B tumors	Human follicular lymphoma (FL) tumors	18 Surface markers and 15 proteins (intracellular proteins)	FFC for phenotyping and BCR signaling proteins of FL B cells and TILs	The abundance of impaired BCR cells in tumor negative correlated with patient overall survival	Predicting tumors relapse and defining new populations of rare and clinically significant cells	Irish et al. (2010)
Brain tumor	GBMs, several low-grade tumor types	4 Intracellular proteins	Image cytometry for 4 proteins within mTOR signaling	Striking inter- and intratumor heterogeneity of GBMs	Molecular diagnostic platform for personalized medicine	Sun et al. (2010)
GBMs	GBMs in human and mouse models	11 Intracellular proteins	SCBC for 11 proteins within mTOR signaling pathway	Drug resistance to mTOR kinase inhibitor appearing 2.5 days after treatment in GBM model	Clinical assessment for more effectively treating GBM patients such as combination therapy	Wei et al. (2016)

of progression-free and overall survival before the immunotherapy (Krieg et al., 2018). Wei et al. characterized single tumor-infiltrating T cells (TILs) with mass cytometry by assaying 33 surface markers and 10 intracellular markers, including non-T cell lineage markers, T cell differentiation and activation markers, and T cell lineage transcription factors. Both anti-CTLA-4 and anti-PD-1 antibodies were found to expand exhausted-like CD8 T cells, and anti-CTLA-4 antibody was found to modulate an $ICOS^{+}$ Th1-like CD4 effector subset for engaging exhausted-like CD8 T cells. Wei et al. discovered the distinct cellular mechanisms of antitumor immune responses induced by CTLA-4 and PD-1 blockade in the preclinical and clinical analysis (Wei et al., 2017). Recently, Spitzer et al. proposed that effective cancer immunotherapy was depended on the system immunity. To validate this hypothesis, they systematically assessed antitumor immune response in lymph node, spleen, blood, and bone marrow of mice with breast cancer by assaying 39 protein parameters in immune cells with mass cytometry, including surface markers and intracellular proteins. The results showed the emergent population of CD4 T cells in peripheral blood predicted system active immunity and that system active immunity was required for tumor rejection (Spitzer et al., 2017).

3.2 Adoptive Cell Transfer

ACT as a promising treatment method for malignant tumors by transferring genetically engineered T cells to cancer patients. For genetic engineering, T cells from peripheral blood are manipulated to express cancer-specific TCRs or CARs directly against cancer cells. Although CAR-T therapy has delivered encouraging results toward advanced melanoma, bladder cancer and lung cancer, clinical outcomes remain highly variable. For this reason, a comprehensive assessment of antitumor immunity before and during the therapy is of great importance to monitor functions and the heterogeneity of CAR-T or TCR-T cells for enhancing the treatment performance.

To address the functional heterogeneity of genetically engineered immune cells, Ma et al. developed a SCBC microchip to profile a panel of 12 secreted cytokines at the single-cell level for quantitative assessment of peripheral mononuclear blood cells (PBMCs) from healthy donors and TCR-T (genetically engineered MART-1) therapy. The single-cell secretome profiling revealed a significantly higher number of functional subsets found in melanoma patient than in health donors (Ma et al., 2011). Furthermore, Ma et al. investigated antitumor immunity of TCR-T cells with genetically engineered MART-1 in melanoma patients by single-cell profiling a total of 19 functional cytokines in 3 melanoma patients before and after TCR-T cells infused into patients. MART-1^{+} T cells were found to vary in cell number, phenotype, and antitumor effect over time during the immunotherapy, indicating that a long-term efficacy of antitumor immunity could not be maintained during the immunotherapy and that the functional degradation of TCR-T cells was closely related to their functional heterogeneity (Ma et al., 2013). Xue et al. used a 16-plexed SCBC chip for profiling a panel of functional proteins of antigen-specific stimulated CD19 CAR-T cells with single-cell precision, including four groups of secreted cytokines: effector, stimulatory, regulator, and inflammatory, such as Granzyme B, TNF-a, and IFN-γ. The single-cell proteomic results validate the ex vivo efficacy of stimulated CD19 CAR-T cells to B cell malignancies and provide an in-depth understanding of the polyfunctional subsets that drive patient responses. The SCBC chip thereby shows the potential as a tool for assessment of CAR-T products before infusion, including therapy efficacy and emergence of cytokine storm (Xue et al., 2017).

4. CONCLUSION AND OUTLOOK

Cancer immunotherapy has been developed for more than one century and now becomes a very promising way for cancer treatment. However, immunotherapy is limited by varied responses of patients, and the low response rate can be attributed to heterogeneity in both cancer cells and immune cells, as well as the complex tumor–immune interaction within the tumor microenvironment. Different from other therapeutic agents for cancer treatment, antitumor activity by immune cells is complex and dynamic, regulated by the immune system and interaction between tumor and immune system. Thus, a comprehensive characterization of immune cells, especially the genetically engineered immune cells, and the tumor–immune interaction within the tumor microenvironment at the single-cell level are crucial for understanding how immunotherapies work in patients and enhancing the clinical performance of immunotherapies (Ma et al., 2013; Spitzer et al., 2017). Technically, high-multiplexing, high-throughput single-cell proteomic technologies are now available to address this problem. These technologies have been developed to resolve the tumor heterogeneity that is believed to be one of major reasons of drug resistance for cancer therapies. For example, Wei et al. used SCBC technology to investigate the in vivo response of a primary glioblastoma (GBM) model to an mTOR kinase inhibitor. The single-cell proteomic assays on 11 intracellular phosphoproteins associated with hyperactivated signaling pathways of GBM lead to identification of effective and ineffective therapies and therapy combinations (Wei et al., 2016). Irish et al. used FFC to characterize the B-cell receptor (BCR) signaling in subpopulations of human follicular lymphoma (FL) cells at single-cell level. Functional phenotypes of lymphoma B cells, TILs, and nontumor B cells within tumor samples were determined based on 15 intracellular proteins (including SFK, SYK, AKT, and ERK) associated with BCR signaling under a total of 12 stimulation conditions by FFC. The abundance of impaired BCR cells in each tumor was found to be anticorrelated with overall survival of patients. In addition, levels of CD40 in B cells and IL-7 in TILs were capable of stratifying patient survival (Irish et al., 2010).

The application of single-cell proteomic technologies into cancer immunotherapy provides comprehensive evaluation of system immunity at the single-cell level for better understanding the cellular heterogeneity of immune cells and cancer cells and developing more efficacious yet safe immunotherapy. However, there is still plenty of room in technical improvement of these single-cell analysis technologies in terms of sensitivity, throughput, multiplexing capability, and data analysis algorithm. For example, it is technologically challenging to track multiplexed protein secretion signatures from a number of live single cells at different time points. The dynamic secretion landscape provides unique information for better understanding of how cells respond and evolve during treatment over time. With increasing exploration of single cell proteomic analysis in the immunotherapy, we expect a more important role it plays in pharmaceutical companies to accelerate cell therapy products development, and in hospitals as a personalized diagnostic method for precision medicine.

ACKNOWLEDGMENTS

Q.S. is supported by National Key Research and Development Program Grant 2016YFC0900200 and National Natural Science Foundation of China Grant 21775103. Y.L. is supported by National Natural Science Foundation of China (Grant No. 21605143), Youth Innovation Promotion Association CAS (Grant No. 2018217), National High Technology Research and Development Program of China (Grant No. 201405003), and funds from Dalian Institute of Chemical Physics (Grant No. SZ201601). L.L. is supported by China Postdoctoral Science Foundation (Grant No. 2017M611274).

REFERENCES

Agasti, S. S., Liong, M., Peterson, V. M., Lee, H., & Weissleder, R. (2012). Photocleavable DNA barcode-antibody conjugates allow sensitive and multiplexed protein analysis in single cells. *Journal of the American Chemical Society*, *134*(45), 18499–18502. https://doi.org/10.1021/ja307689w.

Bailey, R. C., Kwong, G. A., Radu, C. G., Witte, O. N., & Heath, J. R. (2007). DNA-encoded antibody libraries: A unified platform for multiplexed cell sorting and detection of genes and proteins. *Journal of the American Chemical Society*, *129*(7), 1959–1967. https://doi.org/10.1021/ja065930i.

Bandura, D. R., Baranov, V. I., Ornatsky, O. I., Antonov, A., Kinach, R., Lou, X., et al. (2009). Mass cytometry: Technique for real time single cell multitarget immunoassay based on inductively coupled plasma time-of-flight mass spectrometry. *Analytical Chemistry*, *81*(16), 6813–6822. https://doi.org/10.1021/ac901049w.

Bendall, S. C., & Nolan, G. P. (2012). From single cells to deep phenotypes in cancer. *Nature Biotechnology*, *30*(7), 639–647. https://doi.org/10.1038/nbt.2283.

Bendall, S. C., Nolan, G. P., Roederer, M., & Chattopadhyay, P. K. (2012). A deep profiler's guide to cytometry. *Trends in Immunology*, *33*(7), 323–332. https://doi.org/10.1016/j.it.2012.02.010.

Bendall, S. C., Simonds, E. F., Qiu, P., Amir, e.-A. D., Krutzik, P. O., Finck, R., et al. (2011). Single-cell mass cytometry of differential immune and drug responses across a human hematopoietic continuum. *Science*, *332*(6030), 687–696. https://doi.org/10.1126/science.1198704.

Betts, M. R., Nason, M. C., West, S. M., De Rosa, S. C., Migueles, S. A., Abraham, J., et al. (2006). HIV nonprogressors preferentially maintain highly functional HIV-specific CD8 (+) T cells. *Blood*, *107*(12), 4781–4789. https://doi.org/10.1182/blood-2005-12-4818.

Bhagwat, N., & Carpenter, E. L. (2017). Flow cytometric methods for circulating tumor cell isolation and molecular analysis. *Advances in Experimental Medicine and Biology*, *994*, 105–118. https://doi.org/10.1007/978-3-319-55947-6_5.

Bradshaw, E. M., Kent, S. C., Tripuraneni, V., Orban, T., Ploegh, H. L., Hafler, D. A., et al. (2008). Concurrent detection of secreted products from human lymphocytes by microengraving: Cytokines and antigen-reactive antibodies. *Clinical Immunology*, *129*(1), 10–18. https://doi.org/10.1016/j.clim.2008.06.009.

Britten, C. M., Gouttefangeas, C., Welters, M. J. P., Pawelec, G., Koch, S., Ottensmeier, C., et al. (2008). The CIMT-monitoring panel: A two-step approach to harmonize the enumeration of antigen-specific CD8(+) T lymphocytes by structural and functional assays. *Cancer Immunology, Immunotherapy*, *57*(3), 289–302. https://doi.org/10.1007/s00262-007-0378-0.

Clifton, G. T., Litton, J. K., Arrington, K., Ponniah, S., Ibrahim, N. K., Gall, V., et al. (2017). Results of a phase Ib trial of combination immunotherapy with a CD8+T cell eliciting vaccine and trastuzumab in breast cancer patients. *Annals of Surgical Oncology*, *24*(8), 2161–2167. https://doi.org/10.1245/s10434-017-5844-0.

Cordone, I., Masi, S., Carosi, M., Vidiri, A., Marchesi, F., Marino, M., et al. (2016). Brain stereotactic biopsy flow cytometry for central nervous system lymphoma characterization: Advantages and pitfalls. *Journal of Experimental & Clinical Cancer Research*, *35*(1), 128. https://doi.org/10.1186/s13046-016-0404-1.

De Rosa, S. C., Herzenberg, L. A., Herzenberg, L. A., & Roederer, M. (2001). 11-color, 13-parameter flow cytometry: Identification of human naive T cells by phenotype, function, and T-cell receptor diversity. *Nature Medicine*, 7(2), 245–248. https://doi.org/10.1038/84701.

Dillman, R. O. (2011). Cancer immunotherapy. *Cancer Biotherapy & Radiopharmaceuticals*, *26*(1), 1–64. https://doi.org/10.1089/cbr.2010.0902.

Elitas, M., Brower, K., Lu, Y., Chen, J. J., & Fan, R. (2014). A microchip platform for interrogating tumor-macrophage paracrine signaling at the single-cell level. *Lab on a Chip*, *14*(18), 3582–3588. https://doi.org/10.1039/c4lc00676c.

Fan, R., Vermesh, O., Srivastava, A., Yen, B. K. H., Qin, L. D., Ahmad, H., et al. (2008). Integrated barcode chips for rapid, multiplexed analysis of proteins in microliter quantities of blood. *Nature Biotechnology*, *26*(12), 1373–1378. https://doi.org/10.1038/nbt.1507.

Fisher, D. A. C., Malkova, O., Engle, E. K., Miner, C. A., Fulbright, M. C., Behbehani, G. K., et al. (2017). Mass cytometry analysis reveals hyperactive NF Kappa B signaling in myelofibrosis and secondary acute myeloid leukemia. *Leukemia*, *31*(9), 1962–1974. https://doi.org/10.1038/leu.2016.377.

Gaudilliere, B., Fragiadakis, G. K., Bruggner, R. V., Nicolau, M., Finck, R., Tingle, M., et al. (2014). Clinical recovery from surgery correlates with single-cell immune signatures. *Science Translational Medicine*, *6*(255). https://doi.org/10.1126/scitranslmed.3009701.

Han, Q., Bagheri, N., Bradshaw, E. M., Hafler, D. A., Lauffenburger, D. A., & Love, J. C. (2012). Polyfunctional responses by human T cells result from sequential release of cytokines. *Proceedings of the National Academy of Sciences of the United States of America*, *109*(5), 1607–1612. https://doi.org/10.1073/pnas.1117194109.

Herzenberg, L. A., Parks, D., Sahaf, B., Perez, O., Roederer, M., & Herzenberg, L. A. (2002). The history and future of the fluorescence activated cell sorter and flow cytometry: A view from stanford. *Clinical Chemistry*, *48*(10), 1819–1827.

Hodi, F. S., O'Day, S. J., McDermott, D. F., Weber, R. W., Sosman, J. A., Haanen, J. B., et al. (2010). Improved survival with ipilimumab in patients with metastatic melanoma. *New England Journal of Medicine*, *363*(8), 711–723. https://doi.org/10.1056/NEJMoa1003466.

Hughes, A. J., Spelke, D. P., Xu, Z., Kang, C. C., Schaffer, D. V., & Herr, A. E. (2014). Single-cell western blotting. *Nature Methods*, *11*(7), 749–755. https://doi.org/10.1038/nmeth.2992.

Irish, J. M., Myklebust, J. H., Alizadeh, A. A., Houot, R., Sharman, J. P., Czerwinski, D. K., et al. (2010). B-cell signaling networks reveal a negative prognostic human lymphoma cell subset that emerges during tumor progression. *Proceedings of the National Academy of Sciences of the United States of America*, *107*(29), 12747–12754. https://doi.org/10.1073/pnas.1002057107.

Johnson, L. A., Morgan, R. A., Dudley, M. E., Cassard, L., Yang, J. C., Hughes, M. S., et al. (2009). Gene therapy with human and mouse T-cell receptors mediates cancer regression and targets normal tissues expressing cognate antigen. *Blood*, *114*(3), 535–546. https://doi.org/10.1182/blood-2009-03-211714.

Kaiser, Y., Lakshmikanth, T., Chen, Y., Mikes, J., Eklund, A., Brodin, P., et al. (2017). Mass cytometry identifies distinct lung CD4(+) T cell patterns in Lofgren's syndrome and non-Lofgren's syndrome sarcoidosis. *Frontiers in Immunology*, *8*, 1130. https://doi.org/10.3389/fimmu.2017.01130.

Kang, C. C., Lin, J. M., Xu, Z., Kumar, S., & Herr, A. E. (2014). Single-cell western blotting after whole-cell imaging to assess cancer chemotherapeutic response. *Analytical Chemistry*, *86*(20), 10429–10436. https://doi.org/10.1021/ac502932t.

Kazane, S. A., Sok, D., Cho, E. H., Uson, M. L., Kuhn, P., Schultz, P. G., et al. (2012). Site-specific DNA-antibody conjugates for specific and sensitive immuno-PCR. *Proceedings of the National Academy of Sciences of the United States of America*, *109*(10), 3731–3736. https://doi.org/10.1073/pnas.1120682109.

Konry, T., Smolina, I., Yarmush, J. M., Irimia, D., & Yarmush, M. L. (2011). Ultrasensitive detection of low-abundance surface-marker protein using isothermal rolling circle amplification in a microfluidic nanoliter platform. *Small*, 7(3), 395–400. https://doi.org/10.1002/smll.201001620.

Krieg, C., Nowicka, M., Guglietta, S., Schindler, S., Hartmann, F. J., Weber, L. M., et al. (2018). High-dimensional single-cell analysis predicts response to anti-PD-1 immunotherapy. *Nature Medicine*, *24*(2), 144–+. https://doi.org/10.1038/nm.4466.

Krutzik, P. O., Irish, J. M., Nolan, G. P., & Perez, O. D. (2004). Analysis of protein phosphorylation and cellular signaling events by flow cytometry: Techniques and clinical applications. *Clinical Immunology*, *110*(3), 206–221. https://doi.org/10.1016/j.clim.2003.11.009.

Krutzik, P. O., & Nolan, G. P. (2006). Fluorescent cell barcoding in flow cytometry allows high-throughput drug screening and signaling profiling. *Nature Methods*, *3*(5), 361–368. https://doi.org/10.1038/Nmeth872.

Lewis, J. J., Janetzki, S., Schaed, S., Panageas, K. S., Wang, S. Q., Williams, L., et al. (2000). Evaluation of CD8(+) T-cell frequencies by the Elispot assay in healthy individuals and in patients with metastatic melanoma immunized with tyrosinase peptide. *International Journal of Cancer*, *87*(3), 391–398. https://doi.org/10.1002/1097-0215(20000801)87:3<391::AID-IJC13>3.0.CO;2-K.

Liu, A. Y., Roudier, M. P., & True, L. D. (2004). Heterogeneity in primary and metastatic prostate cancer as defined by cell surface CD profile. *American Journal of Pathology*, *165*(5), 1543–1556. https://doi.org/10.1016/S0002-9440(10)63412-8.

Love, J. C., Ronan, J. L., Grotenbreg, G. M., van der Veen, A. G., & Ploegh, H. L. (2006). A microengraving method for rapid selection of single cells producing antigen-specific antibodies. *Nature Biotechnology*, *24*(6), 703–707. https://doi.org/10.1038/nbt1210.

Lu, Y., Chen, J. J., Mu, L. Y., Xue, Q., Wu, Y., Wu, P. H., et al. (2013). High-throughput secretomic analysis of single cells to assess functional cellular heterogeneity. *Analytical Chemistry*, *85*(4), 2548–2556. https://doi.org/10.1021/ac400082e.

Lu, Y., Xue, Q., Eisele, M. R., Sulistijo, E. S., Brower, K., Han, L., et al. (2015). Highly multiplexed profiling of single-cell effector functions reveals deep functional heterogeneity in response to pathogenic ligands. *Proceedings of the National Academy of Sciences of the United States of America*, *112*(7), E607–E615. https://doi.org/10.1073/pnas.1416756112.

Ma, C., Cheung, A. F., Chodon, T., Koya, R. C., Wu, Z. Q., Ng, C., et al. (2013). Multifunctional T-cell analyses to study response and progression in adoptive cell transfer immunotherapy. *Cancer Discovery*, *3*(4), 418–429. https://doi.org/10.1158/2159-8290.CD-12-0383.

Ma, C., Fan, R., Ahmad, H., Shi, Q. H., Comin-Anduix, B., Chodon, T., et al. (2011). A clinical microchip for evaluation of single immune cells reveals high functional heterogeneity in phenotypically similar T cells. *Nature Medicine*, *17*(6), 738–U133. https://doi.org/10.1038/nm.2375.

Marusyk, A., Almendro, V., & Polyak, K. (2012). Intra-tumour heterogeneity: A looking glass for cancer? *Nature Reviews. Cancer*, *12*(5), 323–334. https://doi.org/10.1038/nrc3261.

Morgan, R. A., Dudley, M. E., Wunderlich, J. R., Hughes, M. S., Yang, J. C., Sherry, R. M., et al. (2006). Cancer regression in patients after transfer of genetically engineered lymphocytes. *Science*, *314*(5796), 126–129. https://doi.org/10.1126/science.1129003.

Nair, N., Mei, H. E., Chen, S. Y., Hale, M., Nolan, G. P., Maecker, H. T., et al. (2015). Mass cytometry as a platform for the discovery of cellular biomarkers to guide effective rheumatic disease therapy. *Arthritis Research & Therapy*, *17*, Article no. 127. https://doi.org/10.1186/s13075-015-0644-z.

Obara, W., Sato, F., Takeda, K., Kato, R., Kato, Y., Kanehira, M., et al. (2017). Phase I clinical trial of cell division associated 1 (CDCA1) peptide vaccination for castration resistant prostate cancer. *Cancer Science*, *108*(7), 1452–1457. https://doi.org/10.1111/cas.13278.

O'Gorman, W. E., Kong, D. S., Balboni, I. M., Rudra, P., Bolen, C. R., Ghosh, D., et al. (2017). Mass cytometry identifies a distinct monocyte cytokine signature shared by clinically heterogeneous pediatric SLE patients. *Journal of Autoimmunity*, *81*, 74–89. https://doi.org/10.1016/j.jaut.2017.03.010.

Perfetto, S. P., Chattopadhyay, P. K., & Roederer, M. (2004). Innovation—Seventeen-colour flow cytometry: Unravelling the immune system. *Nature Reviews. Immunology*, *4*(8), 648–U5. https://doi.org/10.1038/nri1416.

Porter, D. L., Hwang, W. T., Frey, N. V., Lacey, S. F., Shaw, P. A., Loren, A. W., et al. (2015). Chimeric antigen receptor T cells persist and induce sustained remissions in relapsed refractory chronic lymphocytic leukemia. *Science Translational Medicine*, 7(303), 303ra139. https://doi.org/10.1126/scitranslmed.aac5415.

Restifo, N. P., Dudley, M. E., & Rosenberg, S. A. (2012). Adoptive immunotherapy for cancer: Harnessing the T cell response. *Nature Reviews. Immunology*, *12*(4), 269–281. https://doi.org/10.1038/nri3191.

Rosenberg, S. A., Restifo, N. P., Yang, J. C., Morgan, R. A., & Dudley, M. E. (2008). Adoptive cell transfer: A clinical path to effective cancer immunotherapy. *Nature Reviews. Cancer*, *8*(4), 299–308. https://doi.org/10.1038/nrc2355.

Shi, Q., Qin, L., Wei, W., Geng, F., Fan, R., Shin, Y. S., et al. (2012). Single-cell proteomic chip for profiling intracellular signaling pathways in single tumor cells. *Proceedings of the National Academy of Sciences of the United States of America*, *109*(2), 419–424. https://doi.org/10.1073/pnas.1110865109.

Shin, Y. S., Ahmad, H., Shi, Q., Kim, H., Pascal, T. A., Fan, R., et al. (2010). Chemistries for patterning robust DNA microbarcodes enable multiplex assays of cytoplasm proteins from single cancer cells. *Chemphyschem*, *11*(14), 3063–3069. https://doi.org/10.1002/cphc.201000528.

Simpson, T. R., Li, F., Montalvo-Ortiz, W., Sepulveda, M. A., Bergerhoff, K., Arce, F., et al. (2013). Fc-dependent depletion of tumor-infiltrating regulatory T cells co-defines the efficacy of anti-CTLA-4 therapy against melanoma. *The Journal of Experimental Medicine*, *210*(9), 1695–1710. https://doi.org/10.1084/jem.20130579.

Sinkala, E., Sollier-Christen, E., Renier, C., Rosas-Canyelles, E., Che, J., Heirich, K., et al. (2017). Profiling protein expression in circulating tumour cells using microfluidic western blotting. *Nature Communications*, *8*, 14622. https://doi.org/10.1038/ncomms14622.

Smith, C., & Cerundolo, V. (2001). Immunotherapy of melanoma. *Immunology*, *104*(1), 1–7. https://doi.org/10.1046/j.1365-2567.2001.01297.x.

Spitzer, M. H., Carmi, Y., Reticker-Flynn, N. E., Kwek, S. S., Madhireddy, D., Martins, M. M., et al. (2017). Systemic immunity is required for effective cancer immunotherapy. *Cell*, *168*(3), 487–+. https://doi.org/10.1016/j.cell.2016.12.022.

Spitzer, M. H., & Nolan, G. P. (2016). Mass cytometry: Single cells, many features. *Cell*, *165*(4), 780–791. https://doi.org/10.1016/j.cell.2016.04.019.

Sun, J., Masterman-Smith, M. D., Graham, N. A., Jiao, J., Mottahedeh, J., Laks, D. R., et al. (2010). A microfluidic platform for systems pathology: Multiparameter single-cell signaling measurements of clinical brain tumor specimens. *Cancer Research*, *70*(15), 6128–6138. https://doi.org/10.1158/0008-5472.CAN-10-0076.

Taylor, R. J., Falconnet, D., Niemisto, A., Ramsey, S. A., Prinz, S., Shmulevich, I., et al. (2009). Dynamic analysis of MAPK signaling using a high-throughput microfluidic single-cell imaging platform. *Proceedings of the National Academy of Sciences of the United States of America*, *106*(10), 3758–3763. https://doi.org/10.1073/pnas.0813416106.

Tentori, A. M., Yamauchi, K. A., & Herr, A. E. (2016). Detection of isoforms differing by a single charge unit in individual cells. *Angewandte Chemie, International Edition in English*, *55*(40), 12431–12435. https://doi.org/10.1002/anie.201606039.

Topalian, S. L., Hodi, F. S., Brahmer, J. R., Gettinger, S. N., Smith, D. C., McDermott, D. F., et al. (2012). Safety, activity, and immune correlates of anti-PD-1 antibody in cancer. *The New England Journal of Medicine*, *366*(26), 2443–2454. https://doi.org/10.1056/NEJMoa1200690.

Torres, A. J., Contento, R. L., Gordo, S., Wucherpfennig, K. W., & Love, J. C. (2013). Functional single-cell analysis of T-cell activation by supported lipid bilayer-tethered ligands on arrays of nanowells. *Lab Chip*, *13*(1), 90–99. https://doi.org/10.1039/c2lc40869d.

Ullal, A. V., Peterson, V., Agasti, S. S., Tuang, S., Juric, D., Castro, C. M., et al. (2014). Cancer cell profiling by barcoding allows multiplexed protein analysis in fine-needle aspirates. *Science Translational Medicine*, *6*(219), 219ra219. https://doi.org/10.1126/scitranslmed.3007361.

Ullal, A. V., & Weissleder, R. (2015). Photocleavable DNA barcoding antibodies for multiplexed protein analysis in single cells. *Methods in Molecular Biology (Clifton, N.J.)*, *1346*, 47–54. https://doi.org/10.1007/978-1-4939-2987-0_4.

Urbanits, S., Griesmacher, A., Hopfinger, G., Stockhammer, G., Karimi, A., Muller, M. M., et al. (2002). FACS analysis-a new and accurate tool in the diagnosis of lymphoma in the cerebrospinal fluid. *Clinica Chimica Acta*, *317*(1–2), 101–107. https://doi.org/10.1016/S0009-8981(01)00771-9.

Varadarajan, N., Julg, B., Yamanaka, Y. J., Chen, H. B., Ogunniyi, A. O., McAndrew, E., et al. (2011). A high-throughput single-cell analysis of human CD8(+) T cell functions reveals discordance for cytokine secretion and cytolysis. *Journal of Clinical Investigation*, *121*(11), 4322–4331. https://doi.org/10.1172/JCI58653.

Verronese, E., Delgado, A., Valladeau-Guilemond, J., Garin, G., Guillemaut, S., Tredan, O., et al. (2016). Immune cell dysfunctions in breast cancer patients detected through whole blood multi-parametric flow cytometry assay. *Oncoimmunology*, *5*(3), e1100791. https://doi.org/10.1080/2162402X.2015.1100791.

Wei, S. C., Levine, J. H., Cogdill, A. P., Zhao, Y., Anang, N. A. S., Andrews, M. C., et al. (2017). Distinct cellular mechanisms underlie anti-CTLA-4 and anti-PD-1 checkpoint blockade. *Cell*, *170*(6), 1120–1133.e17. https://doi.org/10.1016/j.cell.2017.07.024.

Wei, W., Shin, Y. S., Xue, M., Matsutani, T., Masui, K., Yang, H. J., et al. (2016). Single-cell phosphoproteomics resolves adaptive signaling dynamics and informs targeted combination therapy in glioblastoma. *Cancer Cell*, *29*(4), 563–573. https://doi.org/10.1016/j.ccell.2016.03.012.

Xue, Q., Bettini, E., Paczkowski, P., Ng, C., Kaiser, A., McConnell, T., et al. (2017). Single-cell multiplexed cytokine profiling of CD19 CAR-T cells reveals a diverse landscape of polyfunctional antigen-specific response. *Journal for Immunotherapy of Cancer*, *5*, Article no. 85. https://doi.org/10.1186/s40425-017-0293-7.

Xue, Q., Lu, Y., Eisele, M. R., Sulistijo, E. S., Khan, N., Fan, R., et al. (2015). Analysis of single-cell cytokine secretion reveals a role for paracrine signaling in coordinating macrophage responses to TLR4 stimulation. *Science Signaling*, *8*(381), ra59. https://doi.org/10.1126/scisignal.aaa2155.

Xue, M., Wei, W., Su, Y. P., Kim, J., Shin, Y. S., Mai, W. X., et al. (2015). Chemical methods for the simultaneous quantitation of metabolites and proteins from single cells. *Journal of the American Chemical Society*, *137*(12), 4066–4069. https://doi.org/10.1021/jacs.5b00944.

Yamanaka, Y. J., Szeto, G. L., Gierahn, T. M., Forcier, T. L., Benedict, K. F., Brefo, M. S., et al. (2012). Cellular barcodes for efficiently profiling single-cell secretory responses by microengraving. *Analytical Chemistry*, *84*(24), 10531–10536. https://doi.org/10.1021/ac302264q.

Zhou, J., Wu, Y., Lee, S. K., & Fan, R. (2012). High-content single-cell analysis on-chip using a laser microarray scanner. *Lab on a Chip*, *12*(23), 5025–5033. https://doi.org/10.1039/c2lc40309a.

Zrazhevskiy, P., & Gao, X. (2013). Quantum dot imaging platform for single-cell molecular profiling. *Nature Communications*, *4*, 1619. https://doi.org/10.1038/ncomms2635.

Kolibri 360 GmbH
Ferdinand-Jühlke-Straße 7
99095 Erfurt, Deutschland
produktsicherheit@kolibri360.de